MONOGRAPHS ON STATISTICS AI

General Edit

V. Isham, N. Keiding, T. Louis, N. Reid, R. Tibshirani, and H. Tong

1 Stochastic Population Models in Ecology and Epidemiology *M.S. Barlett* (1960)
2 Queues *D.R. Cox and W.L. Smith* (1961)
3 Monte Carlo Methods *J.M. Hammersley and D.C. Handscomb* (1964)
4 The Statistical Analysis of Series of Events *D.R. Cox and P.A.W. Lewis* (1966)
5 Population Genetics *W.J. Ewens* (1969)
6 Probability, Statistics and Time *M.S. Barlett* (1975)
7 Statistical Inference *S.D. Silvey* (1975)
8 The Analysis of Contingency Tables *B.S. Everitt* (1977)
9 Multivariate Analysis in Behavioural Research *A.E. Maxwell* (1977)
10 Stochastic Abundance Models *S. Engen* (1978)
11 Some Basic Theory for Statistical Inference *E.J.G. Pitman* (1979)
12 Point Processes *D.R. Cox and V. Isham* (1980)
13 Identification of Outliers *D.M. Hawkins* (1980)
14 Optimal Design *S.D. Silvey* (1980)
15 Finite Mixture Distributions *B.S. Everitt and D.J. Hand* (1981)
16 Classification *A.D. Gordon* (1981)
17 Distribution-Free Statistical Methods, 2nd edition *J.S. Maritz* (1995)
18 Residuals and Influence in Regression *R.D. Cook and S. Weisberg* (1982)
19 Applications of Queueing Theory, 2nd edition *G.F. Newell* (1982)
20 Risk Theory, 3rd edition *R.E. Beard, T. Pentikäinen and E. Pesonen* (1984)
21 Analysis of Survival Data *D.R. Cox and D. Oakes* (1984)
22 An Introduction to Latent Variable Models *B.S. Everitt* (1984)
23 Bandit Problems *D.A. Berry and B. Fristedt* (1985)
24 Stochastic Modelling and Control *M.H.A. Davis and R. Vinter* (1985)
25 The Statistical Analysis of Composition Data *J. Aitchison* (1986)
26 Density Estimation for Statistics and Data Analysis *B.W. Silverman* (1986)
27 Regression Analysis with Applications *G.B. Wetherill* (1986)
28 Sequential Methods in Statistics, 3rd edition
G.B. Wetherill and K.D. Glazebrook (1986)
29 Tensor Methods in Statistics *P. McCullagh* (1987)
30 Transformation and Weighting in Regression
R.J. Carroll and D. Ruppert (1988)
31 Asymptotic Techniques for Use in Statistics
O.E. Bandorff-Nielsen and D.R. Cox (1989)
32 Analysis of Binary Data, 2nd edition *D.R. Cox and E.J. Snell* (1989)
33 Analysis of Infectious Disease Data *N.G. Becker* (1989)
34 Design and Analysis of Cross-Over Trials *B. Jones and M.G. Kenward* (1989)

35 Empirical Bayes Methods, 2nd edition *J.S. Maritz and T. Lwin* (1989)
36 Symmetric Multivariate and Related Distributions
K.T. Fang, S. Kotz and K.W. Ng (1990)
37 Generalized Linear Models, 2nd edition *P. McCullagh and J.A. Nelder* (1989)
38 Cyclic and Computer Generated Designs, 2nd edition
J.A. John and E.R. Williams (1995)
39 Analog Estimation Methods in Econometrics *C.F. Manski* (1988)
40 Subset Selection in Regression *A.J. Miller* (1990)
41 Analysis of Repeated Measures *M.J. Crowder and D.J. Hand* (1990)
42 Statistical Reasoning with Imprecise Probabilities *P. Walley* (1991)
43 Generalized Additive Models *T.J. Hastie and R.J. Tibshirani* (1990)
44 Inspection Errors for Attributes in Quality Control
N.L. Johnson, S. Kotz and X. Wu (1991)
45 The Analysis of Contingency Tables, 2nd edition *B.S. Everitt* (1992)
46 The Analysis of Quantal Response Data *B.J.T. Morgan* (1992)
47 Longitudinal Data with Serial Correlation—A state-space approach
R.H. Jones (1993)
48 Differential Geometry and Statistics *M.K. Murray and J.W. Rice* (1993)
49 Markov Models and Optimization *M.H.A. Davis* (1993)
50 Networks and Chaos—Statistical and probabilistic aspects
O.E. Barndorff-Nielsen, J.L. Jensen and W.S. Kendall (1993)
51 Number-Theoretic Methods in Statistics *K.-T. Fang and Y. Wang* (1994)
52 Inference and Asymptotics *O.E. Barndorff-Nielsen and D.R. Cox* (1994)
53 Practical Risk Theory for Actuaries
C.D. Daykin, T. Pentikäinen and M. Pesonen (1994)
54 Biplots *J.C. Gower and D.J. Hand* (1996)
55 Predictive Inference—An introduction *S. Geisser* (1993)
56 Model-Free Curve Estimation *M.E. Tarter and M.D. Lock* (1993)
57 An Introduction to the Bootstrap *B. Efron and R.J. Tibshirani* (1993)
58 Nonparametric Regression and Generalized Linear Models
P.J. Green and B.W. Silverman (1994)
59 Multidimensional Scaling *T.F. Cox and M.A.A. Cox* (1994)
60 Kernel Smoothing *M.P. Wand and M.C. Jones* (1995)
61 Statistics for Long Memory Processes *J. Beran* (1995)
62 Nonlinear Models for Repeated Measurement Data
M. Davidian and D.M. Giltinan (1995)
63 Measurement Error in Nonlinear Models
R.J. Carroll, D. Rupert and L.A. Stefanski (1995)
64 Analyzing and Modeling Rank Data *J.J. Marden* (1995)
65 Time Series Models—In econometrics, finance and other fields
D.R. Cox, D.V. Hinkley and O.E. Barndorff-Nielsen (1996)
66 Local Polynomial Modeling and its Applications *J. Fan and I. Gijbels* (1996)
67 Multivariate Dependencies—Models, analysis and interpretation
D.R. Cox and N. Wermuth (1996)

68 Statistical Inference—Based on the likelihood *A. Azzalini* (1996)

69 Bayes and Empirical Bayes Methods for Data Analysis
B.P. Carlin and T.A Louis (1996)

70 Hidden Markov and Other Models for Discrete-Valued Time Series
I.L. Macdonald and W. Zucchini (1997)

71 Statistical Evidence—A likelihood paradigm *R. Royall* (1997)

72 Analysis of Incomplete Multivariate Data *J.L. Schafer* (1997)

73 Multivariate Models and Dependence Concepts *H. Joe* (1997)

74 Theory of Sample Surveys *M.E. Thompson* (1997)

75 Retrial Queues *G. Falin and J.G.C. Templeton* (1997)

76 Theory of Dispersion Models *B. Jørgensen* (1997)

77 Mixed Poisson Processes *J. Grandell* (1997)

78 Variance Components Estimation—Mixed models, methodologies and applications
P.S.R.S. Rao (1997)

79 Bayesian Methods for Finite Population Sampling
G. Meeden and M. Ghosh (1997)

80 Stochastic Geometry—Likelihood and computation
O.E. Barndorff-Nielsen, W.S. Kendall and M.N.M. van Lieshout (1998)

81 Computer-Assisted Analysis of Mixtures and Applications—
Meta-analysis, disease mapping and others *D. Böhning* (1999)

82 Classification, 2nd edition *A.D. Gordon* (1999)

83 Semimartingales and their Statistical Inference *B.L.S. Prakasa Rao* (1999)

84 Statistical Aspects of BSE and vCJD—Models for Epidemics
C.A. Donnelly and N.M. Ferguson (1999)

85 Set-Indexed Martingales *G. Ivanoff and E. Merzbach* (2000)

86 The Theory of the Design of Experiments *D.R. Cox and N. Reid* (2000)

87 Complex Stochastic Systems
O.E. Barndorff-Nielsen, D.R. Cox and C. Klüppelberg (2001)

88 Multidimensional Scaling, 2nd edition *T.F. Cox and M.A.A. Cox* (2001)

89 Algebraic Statistics—Computational Commutative Algebra in Statistics
G. Pistone, E. Riccomagno and H.P. Wynn (2001)

90 Analysis of Time Series Structure—SSA and Related Techniques
N. Golyandina, V. Nekrutkin and A.A. Zhigljavsky (2001)

91 Subjective Probability Models for Lifetimes
Fabio Spizzichino (2001)

92 Empirical Likelihood *Art B. Owen (2001)*

93 Statistics in the 21st Century *Adrian E. Raftery, Martin A. Tanner, and Martin T. Wells (2001)*

94 Accelerated Life Models: Modeling and Statistical Analysis
Vilijandas Bagdonavičius and Mikhail Nikulin (2001)

95 Subset Selection in Regression, Second Edition
Alan Miller (2002)

96 Topics in Modelling of Clustered Data
Marc Aerts, Helena Geys, Geert Molenberghs, and Louise M. Ryan (2002)

97 Components of Variance *D.R. Cox and P.J. Solomon* (2002)

Statistical Evidence

A likelihood paradigm

RICHARD M. ROYALL
Department of Biostatistics
The Johns Hopkins University
Baltimore
USA

CRC Press is an imprint of the
Taylor & Francis Group, an informa business
A CHAPMAN & HALL BOOK

Originally published by Chapman & hall
First edition 1997

Published 2000 by CRC Press
Taylor & Francis Group
6000 Broken Sound Parkway NW, Suite 300
Boca Raton, FL 33487-2742

First issued in paperback 2022

ISBN 13: 978-1-03-247800-5 (pbk)
ISBN 13: 978-0-412-04411-3 (hbk)

DOI: 10.1201/9780203738665

Publisher's Note
The publisher has gone to great lengths to ensure the quality of this reprint but points out that some imperfections in the original copies may be apparent.

Visit the Taylor & Francis Web site at
http://www.taylorandfrancis.com

and the CRC Press Web site at
http://www.crcpress.com

Library of Congress Cataloging-in-Publication Data

Royall, Richard M.
Statistical evidence : a likelihood paradigm / Richard M. Royall.
p. cm. — (Monographs on statistics and applied probability ; 71)
Originally published: London ; New York : Chapman & Hall, 1997.
Includes bibliographical references and index.
ISBN 0-412-04411-0
1. Estimation theory. 2. Probabilities. 3. Mathematical statistics. I. Title. II. Series.
QA276.8.R69 1999
519.5'44—dc21 99-26362
CIP

Library of Congress Card Number 9926362

Contents

Preface

Science looks to statistics for help in interpreting data. Statistics is supposed to provide objective methods for representing scientific data as evidence and for measuring the strength of that evidence. Statistics serves science in other ways as well, contributing to both efficiency and objectivity through theories for the design of experiments and decision-making, for example. But its most important task is to provide objective quantitative alternatives to personal judgement for interpreting the evidence produced by experiments and observational studies. In this role statistics has made fundamental contributions to science.

All is not well, however. Standard statistical methods regularly lead to the misinterpretation of results of scientific studies. The errors are usually quantitative, as when statistical evidence is judged to be stronger or weaker than it really is. But sometimes they are qualitative – sometimes evidence is judged to support one hypothesis over another when the opposite is true. These misinterpretations are not a consequence of scientists misusing statistics. They reflect instead a critical defect in current theories of statistics.

These problems exist because the discipline of statistics has neglected a key question for which it is responsible: when does a given set of observations support one statistical hypothesis over another? That is, when is it right to say that the observations are evidence in favor of one hypothesis *vis-à-vis* another? The answer to this fundamental question has been known for at least a century (Venn, 1876). However, neither the question nor its simple answer is to be found in most modern statistics textbooks. The reason is that for the last half-century statistical theory has been dominated by a decision-making paradigm – since the work of Neyman and Pearson in the 1930s, basic statistical problems have been formulated not in terms of interpreting data as evidence, but in terms of choosing between alternative courses of action. This has led to the current state of affairs in which the dominant (Neyman–Pearson) *theory* views common statistical procedures as decision-making tools, while much of statistical *practice* consists of using the same procedures for a different purpose, namely, interpreting data as evidence. For example, a test of two hypotheses, *A* and *B*, is represented in Neyman–Pearson statistical theory as a procedure for choosing between them. But in applications, when an optimal test chooses B it is often taken to mean that the data are evidence favoring *B* over *A*, an interpretation that can be quite wrong (section 1.7).

Because of the gap between the current theory and the needs of science for objective quantitative methods to represent and interpret statistical evidence, much of what statistics teaches seems surprising, counterintuitive, and even unscientific. For example, statistics discourages the examination and assessment of observations as they are made, penalizing scientists for looking at their data before the study is complete. (This is made explicit in current methodology for analyzing data from clinical trials.) Consider the researcher who examines his first 20 observations, finds that they are almost statistically significant at a given target level, but not quite, and is encouraged by this finding to continue with his promising study. He is encouraged, that is, unless he consults a statistician, who explains that the researcher is misinterpreting his results. His almost significant statistical test shows not that the study is promising, but that it has failed. Statistical hypothesis-testing theory explains that by performing the test the researcher used up his allowable Type I error probability, so that subsequent observations, no matter how consistent and convincing, can never justify a claim of statistical significance at his target level (Cornfield, 1966; this phenomenon will be discussed below in section 5.3). Finding that the early partial results represent evidence that is only fairly strong precludes the possibility that the evidence in the final results might be quite strong. Does this make sense?

That statistical theory is defective is also revealed in the inability of that theory to answer fundamental questions about what the results of routine statistical analyses mean. One of the most widely used statistical tools is the test of significance, and the interpretation that is central to most applications in science is that results that are barely statistically significant at the 5% level represent moderately strong evidence against the hypothesis that is being tested. This interpretation is endorsed by statisticians of the highest authority. However, in Chapter 3 we see that others, of equally high authority, explain that this interpretation is wrong – such results actually represent only weak evidence if the study is small, but quite strong evidence if it is large. Still others agree that the strength of the evidence does indeed depend on the sample size, but they assert that it is greater in a small study. This dispute has no solution within the Neyman–Pearson theory of hypothesis testing because the concept of evidence is missing altogether from that theory – its authors insisted that rejecting one statistical hypothesis in favor of another signifies a decision to act in a certain way, and nothing else. Statistics has a critical defect.

This monograph is about the problem of representing and interpreting statistical data as evidence. Its focus is on the principle that is fundamental to solving that problem, the law of likelihood. Its point is that statistics' neglect of the law of likelihood has left the discipline's foundations defective, and that, as a consequence, statistics has produced a seriously defective methodology. It explains why, despite the well-known defects of that methodology, science has clung to it. The explanation is given in terms of two entirely different formulations of the problem of testing statistical hypotheses: the formulation of Neyman and Pearson, which serves as today's dominant paradigm of frequentist statistical theory; and that of Fisher, which more often guides the use and interpretation of statistical methods in science.

We examine the strengths and the defects in both the Neyman–Pearson and Fisher paradigms, and propose an alternative. The new paradigm, based on the law of likelihood, provides the explicit objective quantitative concept of evidence that is missing from the old ones, while retaining the key feature of the old paradigms that made them so attractive to science, namely, objective measurement and control of the frequency of misleading results. We show that this new paradigm leads to statistical methods that have a compelling rationale and elegant simplicity.

The likelihood paradigm represents a solution to the dilemma that science has faced since the emergence of the modern Bayesian movement in statistics in the 1950s. The Bayesians drew attention to the logical defects and inconsistencies of frequentist statistical methods (of both the Neyman–Pearson and the Fisher schools). But because Bayesian statistics has a central component that is frankly subjective, science has clung to frequentist methods. The new likelihood paradigm resolves the dilemma by generating statistical theory and methods that are based on the same models and concepts of probability as the old frequentist paradigms (and which also provide for the measurement and control of 'error rates'), but which are free of the inconsistencies. We are no longer forced to choose between the illogic of the frequentists and the subjectivity of the Bayesians.

The notes on which this monograph is based have been developed in a theory course for graduate students in biostatistics at Johns Hopkins. However, the work is intended to be accessible to anyone who has some familiarity with probability and basic statistical methods, and who is curious about the concepts and principles, as opposed to the mathematics or the mechanics, of statistics.

In Chapter 1 we introduce the law of likelihood, and examine its rationale and its scope. We show that it is consistent with the demands of intuition, with the mathematics of probability theory, and with the operational requirements of a rule that is to be used by scientists for interpreting observations as evidence. We also test the law by examining some purported counterexamples.

In Chapters 2 and 3 we describe the concepts, theory, and methods that currently dominate statistical practice. Chapter 2 describes the leading theory, which is based on the work of Neyman and Pearson, and which avoids the concept of evidence altogether. This chapter also examines the use of statistical methods generated by Neyman–Pearson theory for a purpose not envisioned by that theory, namely, interpreting statistical data as evidence. It shows how such misuse of the methods can lead to serious errors. Chapter 3 addresses typical applications of statistical methods in scientific inference. In these applications the results of standard statistical calculations are given an interpretation that is quite different from that of Neyman–Pearson theory. We call these applications 'Fisherian' because of R.A. Fisher's influence on them. They employ a concept of evidence, but one that is incompatible with the law of likelihood. We argue that this concept of evidence is invalid and that, as a result, Fisherian statistical methods can and sometimes do lead to invalid interpretations of data generated by scientific experiments and observational studies.

Chapter 4 interprets the Neyman–Pearson methods of Chapter 2 and the Fisherian methods of Chapter 3 as resulting from different paradigms for statistics, different ways of formulating and solving a simple and basic prototypical problem in planning scientific studies and interpreting their results. It explains the problems with both of these approaches in terms of the inadequacy of their underlying paradigms. It proposes an alternative paradigm that captures what is most persuasive about the old ones, but avoids their basic flaw. The old paradigms use probabilities, in the form of error rates, confidence coefficients, etc., in two distinct roles: to measure the chances of unsatisfactory results (e.g. the probability of rejecting a hypothesis when it is true); and to measure the strength of the evidence represented by an observed set of statistical data. The new paradigm, on the other hand, is based on the law of likelihood and that law's implication that representing and interpreting statistical data as evidence should be done in terms of likelihood ratios, not probabilities. Thus the new paradigm uses probabilities at the planning stage, for measuring and controlling the probabilities that weak or misleading evidence will be generated. But the evidential interpretation of statistical data is done entirely in terms of likelihoods.

In Chapter 5 we use the new likelihood paradigm to analyze some familiar peculiarities and paradoxes that have been generated by the old ones, and to correct some of the old paradigms' false teachings. For example, we see that scientists should not be penalized for looking at data before a study is finished. This does not mean that there are not good reasons for the practice of 'blinding' in many experiments. What it does mean is that the act of analyzing today's data does not alter the meaning of the data, and does not limit how tomorrow's observations might be interpreted.

In Chapter 6 we illustrate the use of likelihoods to represent and interpret statistical evidence. These examples use data drawn from the scientific literature as well as from standard methods textbooks. One very important advantage of likelihood functions is that they are most naturally represented, understood, and communicated graphically. In order to see what the data say, we look at graphs of likelihood functions.

Chapter 7 considers the practical problems presented by nuisance parameters. The general problem is that we are simple-minded creatures who have difficulty understanding relationships in many dimensions. Thus we want techniques that will enable us to represent and interpret the evidence about multiple parameters one at a time. When we are using multi-dimensional parametric models, this quite legitimate desire to analyze and to simplify can be satisfied, in general, only by *ad hoc* techniques for eliminating the parameters that are not of immediate interest. These include using marginal, conditional, and profile likelihoods as well as concepts of exact and approximate orthogonality of parameters.

Bayesian statistics is examined briefly in Chapter 8. Our objective here is simply to make clear the main structural and conceptual differences between the probability models that the Bayesians use and the ones that the frequentists use to articulate their very different representations of the discipline of statistics. We emphasize that the likelihood methods advocated here require only the frequentists' probability models, and do not entail Bayesian prior probability distributions for parameters.

CHAPTER 1

The first principle

1.1 Introduction

In this chapter we distinguish between the specific question whose answer we seek and other important statistical questions that are closely related to it. We find the answer to our question in the simplest possible case, where the proper interpretation of statistical evidence is transparent. And we begin to test that answer with respect to intuition, or face-validity; consistency with other aspects of reasoning in the face of uncertainty (specifically, with the way new evidence changes probabilities); and operational consequences. We also examine some of the common examples that have been cited as proof that the answer we advocate is wrong. We observe two general and profound implications of accepting the proposed answer. These suggest that a radical reconstruction of statistical methodology is needed. Finally, to define the concept of statistical evidence more precisely, we illustrate the distinction between degrees of uncertainty, measured by probabilities, and strength of evidence, which is measured by likelihood ratios.

1.2 The law of likelihood

Consider a physician's diagnostic test for the presence or absence of some disease, *D*. Suppose that experience has shown the test to be a good one, rarely producing misleading results. Specifically, the performance of the test is described by the probabilities shown in Table 1.1. The first row shows that when *D* is actually present, the test detects it with probability 0.95, giving an erroneous negative result with probability 0.05. The second row shows that when *D* is absent, the test correctly produces a negative result with probability 0.98, leaving a false positive probability of only 0.02.

Now suppose that a patient, Mr Doe, is given the test. On learning that the result is positive, his physician might draw one of the

Table 1.1 *A physician's diagnostic test for the presence or absence of disease D*

		Test result	
		Positive	Negative
Disease *D*	Present	0.95	0.05
	Absent	0.02	0.98

following conclusions:

1. Mr Doe probably does not have *D*.
2. Mr Doe should be treated for *D*.
3. The test result is evidence that Mr Doe has *D*.

Which, if any, of these conclusions is appropriate? Can any of them be justified? It is easy to see that under the right circumstances all three might be simultaneously correct.

Consider conclusion 1. It can be restated in terms of the probability that Mr Doe has *D*, given the positive test, $\Pr(D|+)$; it says that $\Pr(D|+) < \frac{1}{2}$. Whether this is true or not depends in part on the result (+) and the characteristics of the test (Table 1.1). But it also depends on the prior (before the test) probability of the condition, $\Pr(D)$. Bayes's theorem shows that

$$\Pr(D|+) = \frac{\Pr(+|D)\Pr(D)}{\Pr(+|D)\Pr(D) + \Pr(+|\text{not-}D)\Pr(\text{not-}D)}$$

$$= \frac{0.95\Pr(D)}{0.95\Pr(D) + 0.02(1 - \Pr(D))}.$$

If *D* is a rare disease, so that $\Pr(D)$ is very small, then it will be true that $\Pr(D|+)$ is small and conclusion 1 is correct (as, for example, if $\Pr(D) = 0.001$, so that $\Pr(D|+) = 0.045$). On the other hand, if *D* were more common – say, with a prior probability of $\Pr(D) = 0.20$ – then $\Pr(D|+)$ would be 0.92, and conclusion 1 would be quite wrong. The validity of conclusion 1 depends critically on the prior probability.

Even if conclusion 1 is correct – say, $\Pr(D|+) = 0.045$ – conclusion 2 might also be correct, and the physician might appropriately decide to treat for *D* even though it is unlikely that *D* is present. This might be the case when the treatment is effective if *D* is present but harmless otherwise, and when failure to treat a patient who actually has D is disastrous. But conclusion 2 would be wrong under different assumptions about the risks associated with the treatment, about the consequences of failure to treat when D is actually present, etc. It is clear that to evaluate conclusion 2 we need, in addition to the information required to evaluate conclusion 1, to know what are the various possible actions and what are their consequences in the presence of D and in its absence.

But how about conclusion 3? The rule we will consider implies that it is valid, independently of prior probabilities, and without reference to what actions might be available or their consequences: the positive test result is evidence that Mr Doe has the disease. Furthermore the rule provides an objective numerical measure of the strength of that evidence.

We are concerned here with the interpretation of a certain kind of observation as evidence in relation to a certain kind of hypothesis. The observation is of the form $X = x$, where X is a **random variable** and x is one of the possible values of X. We begin with hypotheses which, like the two in the example of Mr Doe's test, imply definite numerical probabilities for the observation. Later we will consider more general hypotheses.

> *Law of likelihood*: If hypothesis A implies that the probability that a random variable X takes the value x is $p_A(x)$, while hypothesis B implies that the probability is $p_B(x)$, then the observation $X = x$ is evidence supporting A over B if and only if $p_A(x) > p_B(x)$, and the likelihood ratio, $p_A(x)/p_B(x)$, measures the strength of that evidence (Hacking, 1965).

In our example the hypothesis (A) that Mr Doe has disease D implies that a positive test result will occur with probability 0.95, while hypothesis B, that he does not have D, implies that the probability is only 0.02. Thus, according to the law of likelihood, Mr Doe's positive test is evidence supporting A over B, and conclusion 3 is correct.

1.3 Three questions

The physician's three conclusions can be paraphrased as follows:

1. I believe B to be true.
2. I should act as if A were true.
3. This test result (+) is evidence supporting A over B.

These are answers to three generic questions:

1. What do I believe, now that I have this observation?
2. What should I do, now that I have this observation?
3. What does this observation tell me about A versus B? (How should I interpret this observation as evidence regarding A versus B?)

Cox (1958) distinguished between the problem areas represented by questions 2 and 3 and emphasized the importance of the latter:

> Even in problems where a clear-cut decision is the main object, it very often happens that the assessment of losses and prior information is subjective, so that it will help to get clear first the relatively objective matter of what the data say... In some fields, too, it may be argued that one of the main calls for probabilistic statistical methods arises from the need to have agreed rules for assessing strength of evidence.

The third question is the one we want to answer. Although all three are obviously important, we will consider the first two only to clarify the third. It is the third question that is central to the reporting of statistical data in scientific journals. For example, an epidemiologist might investigate the risk of a certain disease among workers exposed to a chemical agent in comparison to the risk among unexposed workers. He produces a data set, and our objective as statisticians is to understand how the data should be presented and interpreted as evidence about the risks. Suppose it has been hypothesized that exposure might be associated with a substantial increase in the risk of the disease. Are these data evidence supporting that hypothesis? If so, how strong is the evidence for, say, a fivefold increase versus no increase? Is this evidence consistent with that found in other studies? If the published report presents clear answers to such questions then it will be helpful to readers who will use this evidence, along with that from other sources, in deciding whether to move for changes in the workplace, whether to do another, larger, study, whether to undertake an investigation to explain how the chemical exposure might lead to the disease, whether to change jobs, etc. The published paper presents the data, along with analyses that make clear its evidential meaning. The readers will then use the evidence to adjust their beliefs and to help them in making decisions.

We will concentrate on hypotheses of a special kind, statistical hypotheses. A **simple statistical hypothesis** is one that completely specifies the probability distribution of an observable random variable. A **composite statistical hypothesis** asserts that the distribution belongs to a specified set of distributions. In our diagnostic example the random variable X represents the outcome of Mr Doe's test, and the two hypotheses about the presence or absence of D imply two simple statistical hypotheses: if D is present then X has the probability distribution given in the first row of Table 1.1, and if D is absent then X has the distribution given in the second row. When the observations are not numerical, as in Mr Doe's test where the outcomes are 'positive' or 'negative', we will usually give them numerical codes such as $1 \equiv$ 'positive' and $0 \equiv$ 'negative'. The random variable will often be vector-valued, i.e. a realization x of X is not a single number, but an ordered set of numbers, as it would be if we observed not only Mr Doe's test result (x_1) but also his blood pressure (x_2) and pulse rate (x_3). Then the observation would be a vector $x = (x_1, x_2, x_3)$.

The reader might have noticed that the law of likelihood, as stated, does not apply to continuous probability distributions. This limitation is not essential, and Exercise 1.1 extends it to continuous distributions. But for now we must see if the law is persuasive in the simple discrete case.

1.4 Towards verification

Why should we accept the law of likelihood? One favorable point is that it seems to be the natural extension, to probabilistic phenomena, of scientists' established form of reasoning in deterministic situations. If A implies that under specified conditions x will be observed, while B implies that under the same conditions something else, not x, will be observed, and if those conditions are created and x is seen, then this observation is evidence supporting A versus B. This is the law of likelihood in the extreme case of $p_A(x) = 1$ and $p_B(x) = 0$. The law simply extends this way of reasoning to say that if x is more probable under hypothesis A than under B, then the occurrence of x is evidence supporting A over B, and the strength of that evidence is determined by how much greater the probability is under A. This seems both objective and fair – the hypothesis that assigned the greater probability to the observation did the better job of predicting what actually happened, so it is better supported by that observation. If the likelihood ratio, $p_A(x)/p_B(x)$, is very large, then hypothesis A did a much better job than B of predicting which value X would take, and the observation $X = x$ is very strong evidence for A versus B.

One crucial test of the law of likelihood is for consistency with the rules of probability theory. There are serious questions about when it is meaningful to speak of the probability that a hypothesis A is true. But there certainly are some situations where hypotheses have probabilities. (For example, if I generate X by drawing balls from one urn or another, and if I choose which urn to draw from by a coin toss, then the hypotheses corresponding to the two urns both have probability 0.5.)

Suppose A and B are hypotheses for which $\Pr(A)/\Pr(B)$ is the probability ratio before X is observed. The elementary rules governing conditional probabilities imply that after $X = x$ is observed, the probability ratio is changed to

$$\frac{\Pr(A|X = x)}{\Pr(B|X = x)} = \frac{\Pr(X = x|A)\Pr(A)}{\Pr(X = x|B)\Pr(B)} = \frac{p_A(x)}{p_B(x)} \frac{\Pr(A)}{\Pr(B)}. \tag{1.1}$$

This shows that the new evidence, that the observed value of the random variable X is x, changes the probability ratio by the factor $p_A(x)/p_B(x)$, precisely in agreement with the law of likelihood. If we use the law then our interpretations of data as evidence will be consistent with the rules of probability theory; we will never claim that an observation is evidence supporting A over B when the effect of that observation, if A and B had probabilities, would be to reduce the probability of A relative to that of B. Furthermore, the factor $p_A(x)/p_B(x)$, that the law uses to measure the strength of the evidence, is precisely the factor by which the observation $X = x$ would change the probability ratio $\Pr(A)/\Pr(B)$.

It is important to be aware that in asking which is better supported, A or B, we are not assuming that one or the other must be true. On this point, we note that equation (1.1) does not require that the two probabilities, $\Pr(A)$ and $\Pr(B)$, sum to one.

Another crucial test of the law of likelihood is operational – does it work? If we use the law to evaluate evidence, will we be led to the truth? Suppose A is actually false and B is true. Can we obtain observations that, according to the law, are evidence for A over B? Certainly. Does this mean that the law is invalid? Certainly not. Evidence, properly interpreted, can be misleading. This must be the case, for otherwise we would be able to determine the truth (with perfect certainty) from any scrap of evidence that it not utterly ambiguous. It is too much to hope that evidence cannot be misleading. However, we might reasonably expect that strong evidence cannot be misleading very often. We might also expect that, as evidence accumulates, it will tend to favor a true hypothesis over a false one more and more strongly. These expectations are met by the concept of evidence embodied in the law of likelihood, as explained below.

Suppose A implies that X has probability distribution $p_A(\cdot)$, while B implies $p_B(\cdot)$. If B is true then when we observe X it is unlikely that we will find strong evidence favoring the false hypothesis A. Specifically, for any given constant $k > 0$,

$$\Pr(p_A(X)/p_B(X) \geq k) \leq 1/k. \tag{1.2}$$

This is because, if S is the set of values of x that produce a likelihood ratio (in favor of A versus B) of at least k, then when B is correct

$$\Pr(S) = \sum_S p_B(x) \leq \sum_S p_A(x)/k \leq 1/k.$$

The first inequality is obtained because, for every x in S, $p_B(x) \leq p_A(x)/k$, and the second because the sum $\sum_S p_A(x)$ is the probability of S when A is correct, which cannot exceed one.

A similar argument can be used to prove a much stronger result: if an unscrupulous researcher sets out deliberately to find evidence supporting his favorite but erroneous hypothesis (A) over his rival's (B), which happens to be correct, by a factor of at least k, then the chances are good that he will be eternally frustrated. Specifically, suppose that he observes a sequence $X_1, X_2, \ldots$ of independent random variables, identically distributed according to $p_B(\cdot)$. He checks after each observation to see whether his accumulated data are 'satisfactory' (likelihood ratio favors A by at least k), stopping and publishing his results only when this occurs. After n observations the likelihood ratio is $\prod_1^n p_A(x_i)/p_B(x_i)$. It is a remarkable fact that the probability that he will be successful is no greater than $1/k$, and this remains true even if the number of observations he can make is limitless. That is, when B is true,

$$\Pr\left(\prod_1^n p_A(X_i)/p_B(X_i) \geq k \text{ for some } n = 1, 2, \ldots\right) \leq 1/k \tag{1.3}$$

(Robbins, 1970).

In a more positive vein, the law of likelihood, together with the law of large numbers, implies that the accumulating evidence represented by observations on a sequence $X_1, X_2, \ldots$ of independent random variables will eventually strongly favor the truth. Specifically, if the X_i are identically distributed according to p_B, and if p_A identifies any other probability distribution, then the likelihood ratio $\prod_1^n p_A(X_i)/p_B(X_i)$ converges to zero with probability one (Exercise 1.3). This means that we can specify any large number k with perfect certainty that our evidence will favor B over A by at least k if only we take enough observations. The truth will appear. It also implies that along with k we can specify any small number $\varepsilon > 0$, then find a sample size n that will ensure that the probability of finding strong evidence (a likelihood ratio of at least k) supporting B over A is at least $1 - \varepsilon$.

1.5 Relativity of evidence

The law of likelihood applies to pairs of hypotheses, telling when a given set of observations is evidence for one versus the other: hypothesis A is better supported than B if A implies a greater probability for the observations than B does. This law represents a concept of evidence that is essentially relative, one that does not apply to a single hypothesis, taken alone. Thus is explains how observations should be interpreted as evidence for A *vis-à-vis* B, but it makes no mention of how those observations should be interpreted as evidence in relation to A alone.

When there are probabilities, $\Pr(A)$ and $\Pr(B)$, for the hypotheses, the law of likelihood implies that an observation $X = x$ that supports A over B increases the relative probability of A, as expression (1.3) shows. This observation does not necessarily increase the absolute probability of A, however. In fact, an observation that supports A over B can reduce the probabilities of both hypotheses. For example, suppose that there is another hypothesis C and that a priori $\Pr(A) = \Pr(B) = \Pr(C) = \frac{1}{3}$. If $p_A(x) = \frac{1}{6}$, $p_B(x) = \frac{1}{12}$, and $p_C(x) = \frac{1}{3}$, then the effect of the observation $X = x$ is to reduce the probability of A and of B while doubling the probability of A relative to that of B. That is, $\Pr(A|X = x) < \Pr(A)$ and $\Pr(B|X = x) < \Pr(B)$, yet

$$\frac{\Pr(A|X = x)}{\Pr(B|X = x)} = 2\frac{\Pr(A)}{\Pr(B)}.$$

The observation is not evidence supporting A taken alone – it is evidence supporting A over B. Likewise, observations can support A over B while increasing both probabilities, and such observations are evidence against B *vis-à-vis* A, but not evidence against B by itself.

Can a valid rule be found that will guide the interpretation of statistical data as evidence relating to a single hypothesis, without reference to an alternative? We will examine two candidates. The first we call the law of improbability. It states that $X = x$ is evidence against A if $p_A(x)$ is small, that is, if A implies that the observation is improbable. The second, which we call the law of changing probability, states that $X = x$ is evidence for or against A according to whether the effect of the observation is to increase or reduce the probability that A is true.

We will argue that neither of these rules represents a satisfactory concept of evidence for scientific discourse, the first because it is wrong, and the second because it is subjective. The first rule, the law of improbability, has had a powerful influence on statistical thinking. It is often cited as the justification for the popular statistical procedures called tests of significance. It will be considered in detail in Chapter 3, where we examine the rationale for tests of significance and argue that the law of improbability is wrong. The second rule, although stated in terms of a single hypothesis, and not referring to any explicit alternative, actually entails both alternative hypotheses and conditions on how prior probability is distributed among the hypotheses. Although it has had little direct impact on statistical thinking, this rule has received much attention from philosophers (Carnap, 1950; Good, 1962; Salmon, 1983).

The law of changing probability says that:

(i) the observation $X = x$ is evidence supporting A if its effect is to increase the probability of A; that is, $X = x$ supports A if $\Pr(A|X = x) > \Pr(A)$; and
(ii) the ratio $\Pr(A|X = x)/\Pr(A)$ measures the strength of the evidence.

In our diagnostic test example, the law of changing probability says that a positive test is evidence supporting the hypotheses that D is present by the factor $\Pr(D|+)/\Pr(D) = \Pr(+|D)/\Pr(+) = 0.95/\Pr(+)$. To calculate this quantity, we must know $\Pr(D)$, in which case $\Pr(+) = 0.95\Pr(D) + 0.02(1 - \Pr(D))$ and

$$\Pr(D|+)/\Pr(D) = r/[r\Pr(\mathrm{D}) + (1 - \Pr(\mathrm{D}))] \qquad (1.4)$$

where $r = 0.95/0.02$ is the likelihood ratio, $\Pr(+|D)/\Pr(+|\text{not-}D)$. Expression (1.4) is a strictly increasing function of r which equals one when $r = 1$. Thus in this case, according to the law of changing probability, the observation $(+)$ supports D over not-D if and only if the likelihood ratio is greater than one. And the greater the likelihood ratio, the stronger the evidence. This conclusion differs from that implied by the law of likelihood only in that the measure of the evidence's strength depends on $\Pr(D)$ as well as on the ratio $\Pr(+|D)/\Pr(+|\text{not-}D)$; a given likelihood ratio in favor of D is interpreted as stronger evidence when $\Pr(D)$ is small than when this probability is large. If you and I hold different initial values for the probability of D, then we will agree that a positive test is evidence for D, but we will disagree about the strength of that evidence.

However, if the possibilities are richer, then the law of changing probability implies that we need not agree even as to the direction of the support, in favor of disease or against it. This is because, although the law of changing probability appears to measure the absolute evidence for or against hypothesis A, not the evidence for A relative to another hypothesis, this measure is in fact strongly dependent not only on what alternatives to A are considered, but also on the way a priori probabilities are distributed over the alternatives.

Suppose we want to evaluate an observation $X = x$ as evidence relating to hypothesis A, for which we know both the a priori probability of A, $\Pr(A)$, and the probability that $X = x$ if A is true, $p_A(x)$. To apply the law of changing probability we must evaluate $\Pr(A|X = x)/\Pr(A) = p_A(x)/\Pr(X = x)$, and the denominator, $\Pr(X = x)$, depends directly on alternatives to A and their a priori probabilities, as well as on the probabilities that $X = x$ under the various alternatives. For example, if there are only three possible hypotheses, A, B, and C, which have respective a priori probabilities $\Pr(A)$, $\Pr(B)$, and $\Pr(C)$, and which imply respective probabilities $p_A(x)$, $p_B(x)$, and $p_C(x)$ for the event $X = x$, then $\Pr(X = x) = p_A(x)\Pr(A) + p_B(x)\Pr(B) + p_C(x)\Pr(C)$. According to the law of changing probability the evidence for A in $X = x$ is

$$\frac{\Pr(A|X = x)}{\Pr(A)} = \frac{p_A(x)}{\Pr(X = x)}$$

$$= \frac{p_A(x)}{p_A(x)\Pr(A) + p_B(x)\Pr(B) + p_C(x)\Pr(C)}.$$

Not only does this quantity depend on the specific alternatives, B and C, that are considered (and the probabilities of $X = x$ under those alternatives), it also depends on how the a priori probability of not-A is divided between B and C. If $\Pr(A)$ is small and if $p_B(x) < p_A(x) < p_C(x)$, then the effect of the observation $X = x$ will be to increase the probability of A if $\Pr(B)$ is large, but to decrease it if $\Pr(C)$ is large. Whereas the law of likelihood measures the support for one hypothesis A relative to a specific alternative B, without regard either to the prior probabilities of the two hypotheses or to what other hypotheses might also be considered, the law of changing probability measures support for A relative to a specific prior probability distribution over A and its alternatives – the alternatives *and* their a priori probabilities are essential to the law of changing probability, although the formula $\Pr(A|X = x)/\Pr(A)$ conceals this dependence.

The law of changing probability is of limited usefulness in scientific discourse because of its dependence on the prior probability distribution, which is generally unknown and/or personal. Although you and I agree (on the basis of the law of likelihood) that given evidence supports A over B, and C over both A and B, we might disagree about whether it is evidence supporting A (on the basis of the law of changing probability) purely on the basis of our different judgements of the a priori probabilities of A, B, and C.

1.6 Strength of evidence

How strong is the evidence when the likelihood ratio is 2? . . . Or 20? Many scientists (and journal editors) are comfortable interpreting a statistical significance level of 0.05 to mean that the observations are 'pretty strong evidence' against the null hypothesis, and a level of 0.01 to mean 'very strong evidence'. Are there reference values of likelihood ratios where corresponding interpretations are appropriate? (Later, in Chapter 3, we will show that these interpretations of significance levels are not appropriate.)

There are two easy ways to develop a quantitative understanding of likelihood ratios. One is to consider some uncomplicated examples where intuition is strong, and examine the likelihood ratios for various imagined observations. The other is to characterize likelihood ratios in terms of their impact on prior probabilities.

1.6.1 A canonical experiment

Suppose we have two identical urns, one containing only white balls, and the other containing equal numbers of white and black balls. One urn is chosen and we draw a succession of balls from it, after each draw returning the ball to the urn and thoroughly mixing the contents. We have two hypotheses about the contents of the chosen urn, 'all white' and 'half white', and the observations are evidence.

Suppose you draw a ball and it is white. Suppose you draw again, and again it is white. If the same thing happens on the third draw, many would characterize these three observations as 'pretty strong' evidence for the 'all white' urn versus the 'half white' one. The likelihood ratio is $2^3 = 8$.

Table 1.2 *Number of successive white balls* (b) *corresponding to values of a likelihood ratio* (LR)

LR	10	20	50	100	1000
b	3.3	4.3	5.6	6.6	10.0

If we observe b successive white balls, then the likelihood ratio in favor of 'all white' over 'half white' equals $1/(\frac{1}{2})^b$, or 2^b. A likelihood ratio of 2 measures the evidence obtained on a single draw when a white ball is observed. If you would consider that observing white balls on each of three draws is 'pretty strong' evidence in favor of 'all white' over 'half white', then a likelihood ratio of 8 is pretty strong evidence.

For interpreting likelihood ratios in other problems it is useful to convert them to hypothetical numbers of white balls (Table 1.2): a likelihood ratio of k corresponds to b white balls, where $k = 2^b$, or $b = \ln k/\ln 2$. Thus if you have observations giving a likelihood ratio $p_A(x)/p_B(x) = 20$, then you have evidence favoring A over B of the same strength as the evidence favoring 'all white' over 'half white' in $b = \ln 20/\ln 2 = 4.3$ consecutive white balls (stronger than four white balls, but not as strong as five). In the diagnostic test of section 1.2 a positive result, with a likelihood ratio of 47.5, is evidence supporting D over not-D of the same strength as that favoring the 'all white' urn when $b = 5.6$ consecutive white balls are drawn.

1.6.2 Effects of likelihood ratios

Some find the preceding statements dubious. To them it is not clear that a likelihood ratio of 4, say, represents the same strength of evidence in all contexts. These doubts come from failure to distinguish between the strength of the evidence, which is constant, and its implications, which vary according to the context of each application (prior beliefs, available actions, etc.).

The key point is that observations with a likelihood ratio of 4 are evidence strong enough to quadruple a prior probability ratio. The values of the prior probabilities do not matter, nor does their ratio. The effect is always the same: a likelihood ratio of 4 produces a fourfold increase in the probability ratio. There are no circumstances where the effect is different – say, where a likelihood ratio of 4 produces a threefold or a fivefold increase. Bayes's formula guarantees this:

$$\frac{\Pr(A|X = x)}{\Pr(B|X = x)} = \frac{p_A(x)}{p_B(x)} \frac{\Pr(A)}{\Pr(B)}.$$

Whether the prior probabilities are known or not makes no difference; their ratio, whatever it might be, would be increased k-fold by observations with a likelihood ratio of $p_A(x)/p_B(x) = k$.

Some people are willing to state probabilities for all sorts of hypotheses, while others find it meaningful to speak of 'the probability that H is true' only for some very special hypotheses, such as those in the urn example, 'all white' and 'half white' when I choose the urn by a coin toss. The numerical value of the likelihood ratio, which is given a precise interpretation in this last situation (via Bayes's theorem), retains that meaning more generally: a likelihood ratio of k corresponds to evidence strong enough to cause a k-fold increase in a prior probability ratio, regardless of whether a prior ratio is actually available in a specific problem or not. The situation is analogous to that in physics where a unit of thermal energy, the BTU, is given a concrete meaning in terms of water – one BTU is that amount of energy required to raise the temperature of one pound of water at 39.2°F by 1°F. But it is meaningful to measure thermal energy in BTUs in rating air conditioners and in other situations where there is no water at 39.2°F to be heated. Likewise the likelihood ratio, given a concrete meaning in terms of prior probabilities, retains that meaning in their absence.

1.7 Counterexamples

We have seen that the law of likelihood is intuitively attractive; that in special situations where we know how to interpret evidence precisely (via its effect on the probabilities of hypotheses), the law is consistent with what we know to be correct; and that it works. We must test it further by examining its implications, but we will first inspect two examples which have convinced some that the law is false. Another purported counterexample is considered in section 1.10.

1.7.1 A trick deck?

I shuffle an ordinary-looking deck of playing cards and turn over the top card. It is the ace of diamonds. According to the law of likelihood, the hypothesis that the deck consists of 52 aces of diamonds

(H_1) is better supported than the hypothesis that the deck is normal (H_N) by the factor $\Pr(A\blacklozenge|H_1)/\Pr(A\blacklozenge|H_N) = 52$. (In comparison with the urn example, this is stronger than the evidence favoring 'all white' over 'half white' when five consecutive draws produce white balls.)

Some find this disturbing. Although the evidence is supposed to be strong, they would not be convinced that there are 52 aces of diamonds instead of a normal deck. Furthermore, it seems unfair; no matter what card is drawn, the law implies that the corresponding trick-deck hypothesis (52 cards just like the one drawn) is better supported than the normal-deck hypothesis. Thus even if the deck is normal we will always claim to have found strong evidence that it is not.

The first point rests on confusing evidence and belief (questions 3 and 1 in section 1.3). If drawing an ace of diamonds does not convince you that the deck has 52 aces of diamonds (H_1), this does not mean that the observation is not strong evidence in favor of H_1 versus H_N. It means simply that the evidence is not strong enough to overcome the prior improbability of H_1 relative to H_N. Edwards (1970) highlighted the role of prior opinion in our reaction to this example by considering how an individual with a somewhat different background might react:

> A Martian faced with this problem would find the first hypothesis [H_1] most appealing; are not all the cards identical in size and shape, with identical patterns on the side exposed to view? How natural, then, that they should all have the same design on the other side.

The interplanetary perspective is not necessary; we can change our own prior beliefs and see how this changes the example. Suppose I show you two decks, one normal and one actually composed of 52 aces of diamonds. I choose a deck by a coin toss, shuffle the chosen deck, and draw one card. It is an ace of diamonds. Now the conclusion that the deck is not the normal one looks quite reasonable. The evidence represented by the ace of diamonds is the same as before; it is the prior probabilities that have changed. Now the beliefs after seeing the evidence are dominated by that evidence, whereas before they were dominated by the prior beliefs.

The second objection to the law, that H_N is treated unfairly, rests on a misinterpretation: 'evidence supporting H_1 over H_N' is not 'evidence against H_N'. Consider the 51 additional different trick-deck hypotheses, $H_2, \ldots, H_{52}$, one stating that all 52 cards are fours of clubs, etc. Observing the ace of diamonds is evidence supporting H_1 over H_N. It is also evidence supporting H_N over H_2, H_3, etc., decisively. It is not evidence for or against H_N alone.

As is often true, a Bayesian calculation can help to clarify the point. Suppose that there is some prior probability π that the deck is normal, and that if it is not normal, then it must be one of the 52 trick decks, all of which are equally probable. Thus $\Pr(H_N) = \pi$ and $\Pr(H_j) = (1 - \pi)/52$, for $j = 1, 2, \ldots, 52$. How are these probabilities changed by the observation of an ace of diamonds? Bayes's theorem reveals that

$$\Pr(H_N|A\blacklozenge) = \pi,$$

$$\Pr(H_1|A\blacklozenge) = 1 - \pi,$$

$$\Pr(H_j|A\blacklozenge) = 0, \qquad j = 2, 3, \ldots, 52.$$

The probability of H_N is unchanged by the observation; the probability of H_1 is increased by a factor of 52, while the probabilities of all the other trick-deck hypotheses are driven to zero. The entire probability, $1 - \pi$, that was distributed evenly over the 52 trick-deck hypotheses is now concentrated on H_1. The probability ratio of H_1 to H_N has increased sharply, from $(1 - \pi)/52\pi$ to $(1 - \pi)/\pi$. But if $\pi > \frac{1}{2}$, this ratio is still less than one and the normal deck remains the more probable.

This example shows the importance, as well as the difficulty, of maintaining the critical distinction between *evidence* and *confidence* (degree of belief). The next example makes a similar point.

1.7.2 Greater confidence without stronger evidence?

Suppose that two distributions, labelled θ_1 and θ_2, both assign the same probability to a specific outcome x – say, $f(x; \theta_1) = f(x; \theta_2) = 1/20$. The hypotheses $H_1\colon \theta = \theta_1$ and $H_2\colon \theta = \theta_2$ both imply that the event $X = x$ has probability 1/20, so that, according to the law of likelihood, the occurrence of this event is not evidence supporting either hypothesis over the other.

Now consider the composite hypothesis: $H_C\colon \theta = \theta_1$ or θ_2. Because this hypothesis also implies that the event $X = x$ has probability 1/20, the same as the probability under H_1, the law asserts that this evidence does not support H_C over H_1 – these two hypotheses are equally well supported. But H_C must be true if H_1 is; therefore H_C is more likely to be true, more plausible, more believable, more tenable than H_1. Does this not imply that the evidence really does support H_C over H_1, contrary to the law?

There are two pieces of evidence here. One is statistical, the observation, $X = x$. The other is logical, the relationship between the two hypotheses. This second bit of evidence implies that H_C is the more credible hypothesis, independently of the first. It does not imply that the statistical evidence supports H_C over H_1. On the other hand, the law of likelihood addresses only the statistical evidence, not that in the logical relationship between the hypotheses. There is no inconsistency in acknowledging both that H_C is more credible than H_1 (because of their logical relationship, and independently of the observation) and that the observation $X = x$ is evidence favoring neither.

1.8 Testing simple hypotheses

According to the law of likelihood, the strength of statistical evidence for one hypothesis *vis-à-vis* another is measured by the likelihood ratio. This ratio also plays a central role in the Neyman–Pearson theory of hypothesis testing, but that theory addresses a different problem than the law of likelihood does. Neyman–Pearson theory, which will be discussed in Chapter 2, is aimed at using the observations to choose between two hypotheses, H_1 and H_2, not at representing or interpreting the observations as evidence. The choice is made as follows. Before X is observed a set R of possible values of X is selected. This set is called the **critical region**. Then if $X = x$ is observed and x is in R, H_2 is chosen; if x is not in R, H_1 is chosen.

Neyman and Pearson (1933) pointed out that two types of error can be made: if H_1 is true then an observation in R will lead to erroneous choice of H_2, a Type I error; a Type II error occurs when H_2 is true but the observation is not in R, so that H_1 is chosen. The probability of a Type I error is called the **size** of the critical region and is denoted by α.

For the case of simple hypotheses, H_1 and H_2, Neyman and Pearson sought, among all critical regions whose size does not exceed a specified value, such as $\alpha \leq 0.05$, the one that has the smallest Type II error probability. They discovered that this best critical region is determined by the likelihood ratio. It is $R = \{x: f_2(x)/f_1(x) \geq k\}$. That is, the best test procedure is 'Choose H_2 if the likelihood ratio is at least k' where k is chosen to give the desired size, α.

For example, suppose the hypotheses specify different values for the success probability in 30 *Bernoulli*(θ) trials – say, $H_1: \theta = \frac{1}{4}$

and H_2: $\theta = \frac{3}{4}$. When the number of successes observed is x, the likelihood ratio in favor of H_2 over H_1 is $f_2(x)/f_1(x) = 3^{2x-30}$. The best critical region with size $\alpha = 0.05$ contains all values of x for which the likelihood ratio is at least $k = 3^{24-30} = \frac{1}{729}$, that is, $x \geq 12$. (Under H_1 the probability of 12 or more successes in 20 trials is only 0.05.)

It is reassuring to find that the best test calls for choosing H_2 when evidence favors H_2 over H_1 by a sufficiently large factor (k). But, as this example shows, the critical factor k can be less than one, and in that case the test sometimes calls for choosing H_2 when the evidence actually favors H_1. For instance, when $x = 12$ is observed, the test calls for choosing H_2, although the observation is strong evidence supporting H_1 over H_2 ($f_1(12)/f_2(12) = 729$, evidence stronger than when nine consecutive white balls are drawn in the urn example of section 1.6).

Similarly, the observations $x = 13$ and 14 are fairly strong evidence in favor of H_1. And the observation $x = 15$, which represents a success rate of $\frac{1}{2}$, equally far from the two hypothesized values, $\theta = \frac{1}{4}$ and $\theta = \frac{3}{4}$, is not evidence supporting H_2 over H_1, but utterly neutral evidence ($f_2(15)/f_1(15) = 1$).

Although likelihood theory and Neyman–Pearson testing theory have much in common, it is clear that there are fundamental differences. While likelihood theory addresses the last of the physician's three questions in Chapter 1 (What does this observation say about H_1 versus H_2?), Neyman–Pearson theory is concerned with the second question (What should I do?). It is interesting to note that the link between the two theories would be even stronger if, instead of minimizing the probability of a Type II error for a fixed value of α, Neyman and Pearson had sought to minimize the sum of the two error probabilities; in that case they would have found that the best critical region consists of those observations whose likelihood ratio is greater than one. That is, they would have found that the best rule is to choose the hypothesis that is better supported by the observations (Exercise 1.6; see Cornfield, 1966). We will take a closer look at the Neyman–Pearson statistical theory in Chapter 2.

1.9 Composite hypotheses

The law of likelihood explains how an observation on a random variable should be interpreted as evidence in relation to two simple statistical hypotheses. It also applies to some composite hypotheses, such as H_C in section 1.7. But it does not apply to composite hypotheses generally. A simple example shows why this is so.

Suppose three probability distributions, labelled θ_1, θ_2, and θ_3, are under consideration for a random variable X. In particular, we want to evaluate an observation $X = x$ as evidence for the simple hypothesis $H_2: \theta = \theta_2$ *vis-à-vis* the composite $H_C: \theta = \theta_1$ or θ_3. Suppose $f(x;\theta_1) > f(x;\theta_2) > f(x;\theta_3)$, that is, the observation is evidence supporting $H_1: \theta = \theta_1$ over H_2, but it also supports H_2 over $H_3: \theta = \theta_3$.

For example, X might be the number of white balls in five draws (with replacement) from an urn whose proportion of white balls is either one-fourth (θ_1), one-half (θ_2) or three-fourths (θ_3). If none of the five draws produces a white ball ($X = 0$), this is evidence supporting H_1 over H_2 by a factor of $(\frac{3}{4})^5/(\frac{1}{2})^5 = \frac{243}{32}$, or about 7.6. But it also supports H_2 over H_3, by $(\frac{1}{2})^5/(\frac{1}{4})^5 = 32$. How about H_C (proportion is either $\frac{1}{4}$ or $\frac{3}{4}$) versus H_2? Because H_C does not imply a definite probability for the observation, the law of likelihood is silent.

Perhaps the law as stated above is unnecessarily restricted, and an acceptable extension might be found which would imply that $X = 0$ is evidence for H_C versus H_2. One argument that might be advanced in support of this speculation is as follows: H_C must be true if H_1 is true, and H_1 is supported over H_2. Thus it might seem reasonable to say that H_C is also supported over H_2. But this argument rests on the same fallacy as the one at the end of section 1.7 – confusing what the logical structure implies with what the statistical data tell us.

Examination of the evidence's effect on the relative probabilities of H_C and H_2, quantified by Bayes's formula, confirms that the suggested extension of the law of likelihood goes too far: if the three values θ_1, θ_2, and θ_3 have respective prior probabilities p_1, p_2, and p_3, then

$$\frac{\Pr(H_C|X=0)}{\Pr(H_2|X=0)} = \frac{(243/32)p_1 + (1/32)p_3}{p_2}.$$

This ratio is larger or smaller than the a priori probability ratio, $(p_1 + p_3)/p_2$, according to whether p_1/p_3 is larger or smaller than 31/211. Observation of no white balls in five draws causes an increase in the probability of H_C, compared to that of H_2, if the ratio of prior probabilities of the components of H_C, p_1/p_3, is large enough, and otherwise causes a decrease.

More generally, let r_{12} be the likelihood ratio of H_1 to H_2, $f(x;\theta_1)/f(x;\theta_2)$, and let r_{32} be the ratio of H_3 to H_2. Then

$$\frac{\Pr(H_C|X=x)}{\Pr(H_2|X=x)} = \frac{r_{12}p_1 + r_{32}p_3}{p_2},$$

which equals

$$\frac{[r_{12}w + r_{32}(1-w)](p_1+p_3)}{p_2},$$

where $w = p_1/(p_1+p_3)$. Thus the term in square brackets is the factor by which the prior probability ratio $\Pr(H_C)/\Pr(H_2)$ is altered by the observation $X = x$. This factor is a weighted average of the two likelihood ratios, r_{12} and r_{32}, the weights being the proportions of the total probability of H_C, $p_1 + p_3$, that are assigned to the respective components, H_1 and H_3. Since this factor is at least as large as the smaller of the two likelihood ratios, we can properly characterize the observation as evidence favoring H_C over H_2 when both r_{12} and r_{32} exceed unity, that is, when the evidence supports each of the components of H_C, H_1 and H_3, over H_2. But when $r_{12} > 1 > r_{32}$ we can interpret the evidence for H_C versus H_2 only in relation to the price probability ratio p_1/p_3.

What, then, can we say about statistical evidence when many more than two simple probability distributions are of genuine interest? Does the law of likelihood (without prior probabilities) provide a means of representing and measuring the evidence which is appropriate for scientific interpretation and communication? To show that it does, we consider an example that is examined in greater detail in section 6.2. In that example medical researchers are interested in the success probability, θ, associated with a new treatment. They are particularly interested in how θ relates to the old treatment's success probability, believed to be about 0.2. They have reason to hope that θ is considerably greater, perhaps 0.8 or even greater. To obtain evidence about θ, they carry out a study in which the new treatment is given to 17 subjects, and find that it is successful in nine.

A standard statistical analysis of their observations would use a *Bernoulli*(θ) statistical model and test the composite hypotheses H_1: $\theta \leq 0.2$ versus H_2: $\theta > 0.2$. That analysis would show that H_1 can be rejected in favor of H_2 at any significance level greater than 0.003, a result that is conventionally taken to mean that the observations are very strong evidence supporting H_2 over H_1.

But because H_1 contains some simple hypotheses that are better supported than some hypotheses in H_2 (e.g. $\theta = 0.2$ is better supported than $\theta = 0.9$ by a likelihood ratio of $LR = (0.2/0.9)^9(0.8/0.1)^8 = 22.2$), the law of likelihood does not allow the characterization of these observations as strong evidence for H_2 over H_1.

What does it allow us to say? One statement that we can make is that the observations are only weak evidence in favor of $\theta = 0.8$ versus $\theta = 0.2$ ($LR = 4$). We can also say that they are rather strong evidence supporting $\theta = 0.5$ over any of the values under $H_1\colon \theta \leq 0.2$ ($LR > 89$), and at least moderately strong evidence for $\theta = 0.5$ over any value $\theta \geq 0.8$ ($LR > 22$). These ratios change very little if we replace $\theta = 0.5$ by slightly different values. Thus we can say that the observation of nine successes in 17 trials is rather strong evidence supporting success rates of about 0.5 over the rate 0.2 that is associated with the old treatment, and at least moderately strong evidence for the intermediate rates versus the rates of 0.8 or greater that we were hoping to achieve.

The law of likelihood does not allow us to characterize the evidence in terms of the hypotheses $H_1\colon \theta \leq 0.2$ and $H_1\colon \theta > 0.2$. It forces us to be more specific, to note and report which values greater than 0.2 are better supported than values of 0.2 or less, for example, and by how much. As we will see later (Figure 1.1), the law of likelihood enables us to see, understand, and communicate the evidence as it pertains not just to two pre-selected hypotheses, but to the totality of possible values of θ.

1.10 Another counterexample

I have written numbers on two cards. On one card I wrote the number $\pi^{-1} = 0.318$ and on the other I wrote the value of a standard normal deviate (recorded to three decimal places). One of the cards is lost, and I am curious about the remaining one, which is in my desk drawer. Is it the card on which I deliberately wrote 0.318, or is it the one with the random number? Here we have two simple hypotheses; one, H_D, states that the number on the card, X, equals 0.318 (with probability one), and the other, H_N, states that X has a standard normal probability distribution. If I open the drawer and observe the value of X, I will have evidence concerning these two hypotheses, and if that value is 0.318, it is evidence supporting H_D over H_N by a very large factor. (Since I rounded the normal deviate, the probability of $X = 0.318$ under H_N is approximately $0.001\phi(0.318) = 1/2637$, where ϕ is the standard normal probability density function. Of course, the probability under H_D is one, so the likelihood ratio in favor of H_D, $P_D(X = 0.318)/P_N(X = 0.318)$, is approximately 2637. This is very strong evidence, having about the same strength as that supporting the hypothesis 'all white' balls versus 'half white' balls in the urn example of section 1.6 when we draw 11 consecutive white balls.

This interpretation of the evidence, guided by the law of likelihood, seems entirely reasonable. I know of no arguments to the contrary. In fact, this example has never been cited as a counterexample to the law. But it bears directly on the next one, which has (Hacking, 1972; Hill, 1973; Cox and Hinkley, 1974, p. 52; see also Birnbaum, 1969, pp. 127–8).

Supposing that X has a normal distribution, consider the evidence in the single observation $X = x$. The likelihood ratio for comparing the evidence for simple hypotheses H_1: $N(\mu_1, \sigma_1^2)$ and H_2: $N(\mu_2, \sigma_2^2)$ is

$$\frac{\sigma_2}{\sigma_1}\exp\frac{1}{2}\left[\left(\frac{x-\mu_2}{\sigma_2}\right)^2 - \left(\frac{x-\mu_1}{\sigma_1}\right)^2\right],$$

which increases without limit as (μ_1, σ_1^2) approach the values $(x, 0)$. This means that, regardless of what the true values of (μ, σ^2) are, so long as σ^2 is not zero, we will always find very strong evidence in favor of another hypothesis, namely that $\mu = x$ and $\sigma^2 = 0$. Uneasiness with this conclusion appears to grow from its misinterpretation, of which there are at least three varieties:

1. The conclusion means that no matter what is observed, it will be strong evidence against the true hypothesis (but this seems both unfair and incorrect).
2. The conclusion means that whatever value x is observed, I should be moved to attach a high degree of belief to the hypothesis $N(x, 0)$, but I am not so moved.
3. The conclusion means that the evidence is strong that σ^2 is very small (but it is intuitively clear that with no prior information about μ, one observation can provide no evidence about σ^2).

The first two misinterpretations are analogous to those already discovered in the earlier example of one card drawn from a well-shuffled deck. The error involved in the first is that it overlooks the relativity of evidence: the fact that we can find some other hypothesis that is better supported than H does not mean that the observations are evidence against H. Reaction 2 was also explained in the same example: evidence must not be confused with belief. Whether or not I am moved to attach a high degree of belief to $N(x, 0)$ depends on my prior belief in that hypothesis. If I remain skeptical, it does not show that the evidence does not favor $N(x, 0)$ over every other hypothesis. What it does show is that my prior skepticism was so strong that it is not overwhelmed by this evidence. As the previous example with $x = 0.318$ showed, if $N(x, 0)$ had been a hypothesis that was plausible before X was observed, then the observation $X = x$ would have elicited a high degree of confidence in its truth.

The third interpretation, that $X = x$ is strong evidence that σ^2 is small, has many facets. One problem is that 'σ^2 is small', or even the more restrictive statement '$\sigma^2 = 0$', is a composite hypothesis, allowing the other parameter μ to range over the entire real line. Concerning the relative support for the composite hypotheses, $\sigma^2 = 0$, versus $\sigma^2 = \sigma_0^2$, say, the law is silent. A claim that the former is the better supported must rest on some additional principle or convention; it is not sanctioned by the law of likelihood. The problem of evaluating evidence concerning one parameter in models that also contain other parameters ('nuisance' parameters) is one with no general solution. The cause has already been described in section 1.9. We address this problem in Chapters 6 and 7, where we will see that for most problems there are quite satisfactory *ad hoc* methods for representing and interpreting evidence in the presence of nuisance parameters.

1.11 Irrelevance of the sample space

The law of likelihood says that the evidence in an observation, $X = x$, as it pertains to two probability distributions labelled θ_1 and θ_2, is represented by the likelihood ratio, $f(x; \theta_1)/f(x; \theta_2)$. In particular, the law implies that for interpreting the observation as evidence for hypothesis H_1: $\theta = \theta_1$ *vis-à-vis* H_2: $\theta = \theta_2$, only the likelihood ratio is relevant. What other values of X might have been observed, and how the two distributions in question spread their remaining probability over the unobserved values is irrelevant – all that counts is the ratio of the probabilities of the observation under the two hypotheses.

This is made clear by examples like the following (Pratt, 1961). Suppose that the hypotheses concern the probability θ of heads when a particular bent coin is tossed, and that, to generate evidence about θ, 20 tosses are made. The result is reported in code, and you, knowing the code, will learn precisely how many tosses produced heads. I know only the code-word for '6', so that from the report I can determine only whether the outcome is '6' or 'not-6'. Thus you will observe a random variable X taking values $x = 0, 1, \ldots, 20$ with probabilities $\binom{20}{x}\theta^x(1-\theta)^{20-x}$. I will observe Y, which equals 6 with probability $\binom{20}{6}\theta^6(1-\theta)^{14}$ and some other value, c, representing the outcome 'not-6', with probability $1 - \binom{20}{6}\theta^6(1-\theta)^{14}$. Your sample space consists of the 21 points $\{0, 1, \ldots, 20\}$ while mine consists of $\{6, c\}$.

Now suppose the experiment is done and heads occur on six tosses. For any probabilities θ_1 and θ_2, my evidence concerning $H_1: \theta = \theta_1$ *vis-à-vis* $H_2: \theta = \theta_2$ is the same as yours; the likelihood ratios for $Y = 6$ and for $X = 6$ are identical: $\theta_1^6(1-\theta_1)^{14}/\theta_2^6(1-\theta_2)^{14}$. Of course, if the experiment had a different outcome, your evidence and mine would have been different; if there had been four heads you would have observed $X = 4$, and your likelihood ratio would have been $\theta_1^4(1-\theta_1)^{16}/\theta_2^4(1-\theta_2)^{16}$, while I would have observed only $Y = c$, giving a likelihood ratio of $[1 - 38\,760\theta_1^6(1-\theta_1)^{14}]/[1 - 38\,760\theta_2^6(1-\theta_2)^{14}]$, but that is irrelevant to the interpretation of the observation at hand, $X = Y = 6$. Although the scientific community might reasonably have chosen to subscribe to your newsletter in preference to mine, on the grounds that you could promise to provide a more detailed description of the observation under most circumstances, for the result that actually occurred, six heads, my report is equivalent, as evidence about θ, to yours. Any concept or technique for evaluating observations as evidence that denies this equivalence, attaching a different measure of 'significance' to your report of this result than to mine, is invalid. Whatever can be properly inferred about θ from your report can be inferred from mine and vice versa. The difference between our sample spaces is irrelevant.

We will see this example again in the following section and in section 3.4, where we use it to illustrate the problems that arise when significance tests are used to measure the strength of statistical evidence. The 'irrelevance of the sample space' is a critically important concept, for it implies a structural flaw that is not limited to significance tests, but pervades all of today's dominant statistical methodology.

1.12 The likelihood principle

Suppose two simple hypotheses for the distribution of a random variable X assign respective probabilities $f_1(x)$ and $f_2(x)$ to the outcome $X = x$, while two different hypotheses for the distribution of another random variable Y assign respective probabilities $g_1(y)$ and $g_2(y)$ to the outcome $Y = y$. If $f_1(x)/f_2(x) = g_1(y)/g_2(y)$ then the evidence in the observation $X = x$ regarding f_1 *vis-à-vis* f_2 is equivalent to that in $Y = y$ regarding g_1 *vis-à-vis* g_2. If a third distribution, f_3, is considered for X, and a third, g_3, for Y, then the two outcomes, $X = x$ and $Y = y$, are equivalent evidence concerning the respective collections of distributions, $\{f_1, f_2, f_3\}$ and $\{g_1, g_2, g_3\}$, if all of the corresponding likelihood ratios are equal: $f_1(x)/f_2(x) = g_1(y)/g_2(y), f_1(x)/f_3(x) = g_1(y)/g_3(y)$, etc. This fact is called the **likelihood principle**; it is usually stated in terms of likelihood functions, which we now define.

It is often convenient to use a parameter θ to label the individual members of a collection of probability distributions, so that each distribution is identified by a unique value of θ. The collection of distributions is $\{f(\cdot;\theta); \theta \in \Theta\}$, where Θ is simply the set of all values of θ. If $\theta = \theta_1$ then the probability that $X = x$ is given by $f(x;\theta_1)$. If the distributions are continuous the same notation, $f(x;\theta_1)$, represents the probability density function at the point x when the distribution is the one labelled θ_1. For a fixed value x, $f(x;\theta)$ can be viewed as a function of the variable θ and it is then called the **likelihood function**. We will use the notation $L(\theta; x)$ for the likelihood function when the value of x needs to be made explicit, and use simply $L(\theta)$ when it does not. The law of likelihood gives this function its meaning: if $L(\theta_1; x) > L(\theta_2; x)$, then the observation x is evidence supporting the hypothesis that θ is θ_1 (that is, the hypothesis that X has the distribution identified with the parameter value θ_1) over the hypothesis that θ is θ_2, and the **likelihood ratio** $L(\theta_1; x)/L(\theta_2; x) \equiv f(x;\theta_1)/f(x;\theta_2)$ measures the strength of that evidence. Because only ratios of its values are meaningful, the likelihood function is defined only up to an arbitrary multiplicative constant – $L(\theta; x) = cf(x;\theta)$.

The likelihood principle asserts that two observations that generate identical likelihood functions are equivalent as evidence; in

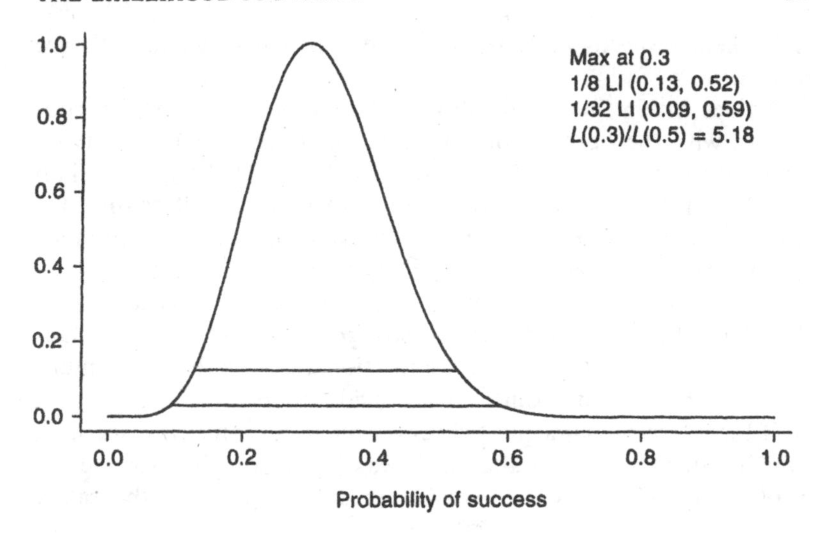

Figure 1.1 *Likelihood for probability of success: six successes observed in 20 trials.*

Birnbaum's (1962) words, 'the "evidential meaning" of experimental results is characterized fully by the likelihood function'.

The example in the previous section concerns a family that contains many more than three distributions. Your observation of the number of heads in 20 tosses of a bent coin was modelled as an observation on a random variable X with a binomial probability distribution, $Bin(20, \theta)$. The probability of six successes is $\Pr(X = 6) = \binom{20}{6}\theta^6(1-\theta)^{14}$, so the likelihood function, $L(\theta)$, is proportional to $\theta^6(1-\theta)^{14}$, for $0 \leq \theta \leq 1$. This function appears in Figure 1.1, which shows that its maximum is at $\theta = 6/20 = 0.3$ (the 'best-supported' hypothesis) and that $\theta = 0.3$ is better supported than $\theta = 0.5$ by a modest factor of $L(0.3)/L(0.5) = (0.3)^6(0.7)^{14}/(0.5)^{20} = 5.2$. This is slightly stronger than the evidence in favor of the 'all white' urn (section 1.6) when two white balls are drawn.

A horizontal line is drawn in Figure 1.1 to show the values of θ where the ratio of $L(\theta)$ to the maximum, $L(0.3)$, is greater than 1/8. Another line shows where it is greater than 1/32. These lines define 'likelihood intervals' (LIs) which, along with the maximizing value, provide a useful summary of what the data say under the present model. The values 1/8 and 1/32 are used because they correspond to the likelihood ratio in the urn example of section

1.6 when, respectively, three and five consecutive white balls are drawn.

Values within the 1/8 likelihood interval are those that are 'consistent with the observations' in the strong sense that there is no alternative value that is better supported by a factor greater than 8. Thus if θ is in this interval, then there is no alternative for which these observations represent 'fairly strong evidence' in favor of that alternative *vis-à-vis* θ. For any θ outside this interval there is at least one alternative value, namely the best-supported value, 0.3, that is better supported by a factor greater than 8. The 1/32 likelihood interval has the same interpretation, but with the bench-mark value 32, representing 'quite strong' evidence, replacing 8.

If I perform the same physical experiment but can only discern Y, which indicates whether the result was '6' or not, then for me the probability of the same observation, six successes, is the same, $\Pr(Y=6)=\Pr(X=6)=\binom{20}{6}\theta^6(1-\theta)^{14}$, so that my likelihood function is the same as yours. The evidence about the probability of heads is represented by that likelihood function (Figure 1.1), and it is the same in both cases – the difference between our sample spaces is irrelevant.

Suppose now that I perform an entirely different experiment. Instead of fixing the number of tosses at 20 I resolve to keep tossing until I get six heads, then to stop. The random variable is now Z, representing the number of tosses required. If I observe $Z=20$, the probability is $\Pr(Z=20)=\binom{19}{5}\theta^6(1-\theta)^{14}$, which is different from $\Pr(X=6)=\Pr(Y=6)$. But the likelihood function is the same, proportional to $\theta^6(1-\theta)^{14}$. For every pair of values, θ_1 and θ_2, the likelihood ratio is the same when $Z=20$ as when $X=Y=6$, so the results are all equivalent as evidence about θ. It is again represented in Figure 1.1 and has precisely the same interpretation in all three cases.

This conclusion calls into question analyses that use the most popular concepts and techniques in applied statistics (unbiased estimation, confidence intervals, significance tests, etc.) when these analyses purport to represent 'what the data say about θ', i.e. to convey the meaning of the observations as evidence about θ. These conventional analyses are questionable because they all certify that there are important differences between observing $X=6$, $Y=6$, and $Z=6$, whereas these three results are in fact evidentially equivalent. For instance, the observed proportion of successes is an unbiased estimator of θ in the first case, but not in the third; a 95% confidence interval for θ based on observation $X = 6$ will differ from those based on $Y = 6$ or $Z = 20$; and for testing hypotheses about θ, e.g. H_1: $\theta \geq 0.5$ versus H_2: $\theta < 0.5$, the three observations will give different p-values. We will consider this example again in Chapter 3.

This is not to say that there are not important differences between the three experiments. The first two were sure to be finished after 20 tosses, while the third could have dragged on indefinitely. The first might have produced 20 consecutive heads, giving a likelihood ratio favoring $\theta_1 = 3/4$ over $\theta_2 = 1/4$ by a factor of more than 3 billion! The second and third experiments cannot possibly generate such strong evidence in favor of θ_1 over θ_2. But the third, by producing a very large value for Z, could have provided much stronger evidence in favor of θ_2 over θ_1 than any possible outcome of the first. The experiments are certainly not equivalent; yet if the first one produces $X = 6$, this outcome is equivalent, as evidence about θ, to the outcome $Y = 6$ of the second and to the outcome $Z = 20$ of the third.

The foregoing conclusion applies even when the parameter θ has a totally different meaning in the different experiments. If you make 20 tosses of a bent coin and observe $X = 6$ heads and I count the number of cars before the sixth Ford has passed my house and observe $Z = 20$ then your evidence about the probability of heads and mine about the probability of a Ford are equivalent. Of course, this equivalence only applies to the specific families of probability distribution being considered:

$$\Pr(X = x) = \binom{20}{x}\theta^x(1-\theta)^{20-x}, \quad x = 0, 1, \ldots, 10,$$

for X and

$$\Pr(Z = z) = \binom{z-1}{5}\theta^6(1-\theta)^{z-6}, \quad z = 6, 7, \ldots,$$

for Z. The evidence supports $\theta = 1/4$ over $\theta = 1/10$ by the factor 19.02, regardless of whether it is your evidence and θ is the probability of heads or it is mine and θ is the probability of Fords. The evidential equivalence of your observation and mine *vis-à-vis* our respective families of distributions applies only to comparisons made within those specific families; it clearly does not assume or imply that my family is as adequate as a model for the frequency of Fords as yours is for the frequency of heads.

This concept – that results of two different experiments have the same evidential meaning if they generate the same likelihood function – has been the focus of much controversy among statisticians. Birnbaum (1962) gave a formal statement of the concept, which he called the likelihood principle. Other authors, notably Fisher (1922) and Barnard (1949), had previously promoted the concept, but most statisticians were not convinced of its validity. Birnbaum increased the pressure on the doubters by showing that the likelihood principle can be deduced from two other principles that most of them did find compelling, the sufficiency and conditionality principles. Since the publication of Birnbaum's result in 1962 statistics has struggled to understand it and to resolve the dilemma that it created (Birnbaum, 1962; 1972; 1977; Durbin, 1970; Savage, 1970; Kalbfleisch, 1975; Berger and Wolpert, 1988; Berger, 1990; Joshi, 1990; Bjornstad, 1996).

1.13 Evidence and uncertainty

We have suggested that the concept of statistical evidence is properly expressed in the law of likelihood and that the likelihood function is the appropriate mathematical representation of statistical evidence. Many likelihood functions, like the one in Figure 1.1, for example, look like probability density functions. However, there are critical differences between the two kinds of function, both in terms of what they mean and in terms of what mathematical operations make sense.

Probabilities measure uncertainty and likelihood ratios measure evidence. A probability density function represents the uncertainty about the value of a random variable; it describes how the uncertainty is distributed over the possible values of the variable (the sample space). That uncertainty disappears when the observation is made – then the value of the random variable is known, and that value is evidence about the probability distribution. The likelihood function represents this evidence; it describes the support ratio for any pair of distributions in the probability model.

Sometimes one variable appears in both aspects of a problem. It is itself a potentially observable random variable, and it is also a parameter that identifies the probability distribution of a second random variable. If (X, Y) are random variables with a given joint probability distribution, then after $X = x$ is observed, $f_{Y|X}(y|x)$ represents the uncertainty about the value of Y. (Note that if we denote the second random variable by Θ instead of Y, then we have the Bayesian statistical model and the Bayesian 'solution' to the problem of statistical inference, $f_{\Theta|X}(\theta|x)$. Bayesian statistics will be discussed in Chapter 8.) But the unobserved value y plays the role of a parameter in $f_{X|Y}(x|y)$, so that the observation $X = x$ is statistical evidence about y, generating a likelihood function $L(y) \propto f_{X|Y}(x|y)$ which represents that evidence. Comparing these two functions in a familiar example helps to clarify their differences.

Consider the case when x and y are realizations of random variables, X and Y, having a bivariate normal probability distribution with expected values μ_x and μ_y, variances σ_x^2 and σ_y^2, and covariance σ_{xy}. Suppose the values of all five parameters are known. If x and y have not yet been observed then the uncertainty about the value y is expressed in the marginal probability distribution of Y, which is $N(\mu_y, \sigma_y^2)$. The observation $X = x$ represents evidence about y. It changes the uncertainty, which, after $X = x$ is observed, is represented by the conditional probability distribution of Y, $N(\mu_y + \sigma_{xy}(x - \mu_x)/\sigma_x^2, \sigma_y^2(1 - \rho^2))$, where ρ denotes the correlation coefficient, $\sigma_{xy}/\sigma_x\sigma_y$. This probability density function, $f_{Y|X}(y|x)$, represents the uncertainty about what value, y, of Y will be observed, now that it is known that $X = x$.

On the other hand, the variable y indexes a family of possible probability distributions for X. These are the conditional distributions of X, given $Y = y$, which are $N(\mu_x + \sigma_{xy}(y - \mu_y)/\sigma_y^2, \sigma_x^2(1 - \rho^2))$. Here y has the role of a parameter – each value of y determines a different probability distribution for X, $f_{X|Y}(x|y)$. Thus the observation $X = x$ generates a likelihood function for y,

$$L(y) \propto \exp\{-\tfrac{1}{2}[x - \mu_x - \sigma_{xy}(y - \mu_y)/\sigma_y^2]^2/\sigma_x^2(1 - \rho^2)\}. \quad (1.5)$$

The only variable in this expression is y – everything else, x, μ_x, etc., is fixed at its known value. The ratio of values of this function at any two points y_1 and y_2, $L(y_1)/L(y_2)$, measures the relative support for these two values of the unknown variable y.

If X and Y are independent, so that $\sigma_{xy} = 0$, then the likelihood function (1.5) for y is a constant, indicating that $X = x$ represents no evidence at all about y. Every likelihood ratio $L(y_1)/L(y_2)$ equals one – when $\sigma_{xy} = 0$ no possible value of y is better supported than any other by the observation $X = x$, regardless of the value of x.

When X and Y are not independent the likelihood function is shaped like a normal probability density function centered at the point $\mu_y + \sigma_y^2(x - \mu_x)/\sigma_{xy}$ and with variance $\sigma_y^2(1 - \rho^2)/\rho^2$. That is, the likelihood function given in expression (1.5) can be rewritten in the form:

$$L(y) \propto \exp\left\{-\frac{1}{2}\frac{[y-\mu_y-\sigma_y^2(x-\mu_x)/\sigma_{xy}]^2}{\sigma_y^2(1-\rho^2)/\rho^2}\right\}.$$

This function represents the *evidence* about y in the observation $X = x$. It does not represent the *uncertainty* about y, which is now given by the conditional probability density function of Y, given $X = x$:

$$f_{Y|X}(y|x) \propto \exp\left\{-\frac{1}{2}\frac{[y-\mu_y-\sigma_{xy}(x-\mu_x)/\sigma_x^2]^2}{\sigma_y^2(1-\rho^2)}\right\}. \tag{1.6}$$

This density function is obtained by adjusting the original $N(\mu_y, \sigma_y^2)$ density function, $f_Y(y)$, in the light of the evidence $X = x$. The adjustment is made simply by taking the product of the original density and the likelihood function $L(y)$. To use this density function we must scale it so that its integral over the entire real line equals one. When we do that, by dividing expression (1.6) by $[2\pi\sigma_y^2(1-\rho^2)]^{1/2}$, integration over any interval then gives the probability that the value of Y will fall inside that interval. This implies, for example, that the probability is 0.95 that y will be found in the predictive interval

$$\mu_y + \sigma_{xy}(x-\mu_x)/\sigma_x^2 \pm 1.96\sigma_y(1-\rho^2)^{1/2}. \tag{1.7}$$

On the other hand, the $1/k$ likelihood interval, $\{y; L(y)/\max L(y) \geq 1/k\}$, which is the set of y values such that no alternative is better supported by a likelihood ratio greater than k, is

$$\mu_y + \sigma_y^2(x-\mu_x)/\sigma_{xy} \pm (2\ln k)^{1/2}(\sigma_y/\rho)(1-\rho^2)^{1/2}. \tag{1.8}$$

The probability is 0.95 that the value of the random variable Y associated with the observed value x will fall in the predictive interval (1.7). No such simple probability statement can be made about the likelihood interval (1.8). But that interval can be interpreted as a confidence interval. Suppose we use the value $k = \exp\{(1.96)^2/2\} = 6.83$, so that the coefficient $(2\ln k)^{1/2}$ equals 1.96. Then for any fixed value y of the random variable Y, if we observe the value of the random variable X and construct the interval (1.8), the probability that this random interval will contain y equals 0.95 (Exercise 1.7). The purpose here is not to suggest that likelihood intervals should be interpreted as confidence intervals, but simply to clarify the distinction between the state of uncertainty about y after observing $X = x$, which is represented by the conditional probability density function, and the evidence about y in the observation $X = x$, which is represented by the likelihood function. The distinction is essentially the same as that between the physician's first and third questions in section 1.3, here rephrased as 'What is the state of uncertainty about y, now that we know that $X = x$?' and 'What does the observation $X = x$ tell us about y?'. The probability density function $f_{Y|X}(y|x)$ answers the first question, and the likelihood function $L(y)$ answers the second. When $\sigma_{XY} = 0$, $X = x$ tells us nothing about y. This is properly represented by the flat likelihood function, $L(y) = \text{constant}$; the probability density function, $f_{Y|X}(y|x) \propto \exp\{-\frac{1}{2}(y - \mu_y)^2/\sigma_y^2\}$, represents something quite different.

1.14 Summary

The question that is at the heart of statistical inference – 'When is a given set of observations evidence supporting one hypothesized probability distribution over another?' – is answered by the law of likelihood. This law effectively defines the concept of statistical evidence to be relative, that is, a concept that applies to one distribution only in comparison to another. It measures the evidence with respect to a pair of distributions by their likelihood ratio.

The law of likelihood is intuitively reasonable, consistent with the rules of probability theory, and empirically meaningful. It is, however, incompatible with today's dominant statistical theory and methodology, which do not conform to the law's general implications, the irrelevance of the sample space and the likelihood principle, and which are articulated in terms of probabilities, which measure uncertainty, rather than likelihood ratios, which measure evidence.

Exercises

1.1 The law of likelihood is stated in section 1.2 for discrete distributions. Suppose that two hypotheses, A and B, both imply that a random variable X has a continuous probability distribution, and that these distributions have continuous density functions, $p_A(x)$ and $p_B(x)$ respectively. Can the law be extended to this case? Explain.

1.2 Suppose A implies that X has probability mass (or density) function $p_A(x)$, while B implies $p_B(x)$. When A is true, observing a value of X that represents strong evidence in favor of B ($p_B(x)/p_A(x) \geq k$) is clearly undesirable. We showed in section 1.4 that the probability of this event is small ($\leq 1/k$).

(a) What can you say about the probability of the desirable event, namely, finding strong evidence in favor of A?

(b) It would be nice if, when A is true, the probability of obtaining strong evidence in favor of A is always at least as great as the probability of (misleading) strong evidence in favor of B, that is, if

$$\Pr_A(p_B(X)/p_A(X) \geq k) \leq \Pr_A(p_A(X)/p_B(X) \geq k).$$

Give a simple example to show that this inequality need not hold.

1.3 Prove the result stated at the end of section 1.4, namely, if A implies that $X_1, X_2, \ldots$ are independent and identically distributed with probability mass function $p_A(x)$, while B implies the same, but with a different mass function given by $p_B(x)$, then when B is true the likelihood ratio in favor of A converges to zero with probability one.

1.4 Suppose $X_1, \ldots, X_n$ are independent, identically distributed random variables with a $N(\theta, \sigma^2)$ probability distribution, with σ^2 known. Consider two hypotheses, $H_0\colon \theta = 0$ and $H_1\colon \theta = \theta_1$, where $\theta_1 > 0$. If a sample is observed with $\sqrt{n}\bar{x}/\sigma = 1.645$, then the p-value, or 'attained significance level', for testing H_0 versus H_1 is 0.05. This p-value is often interpreted as meaning that the observations represent fairly strong evidence against H_0. (This interpretation will be discussed later, in section 3.4.) According to the law of likelihood the strength of the evidence depends on the value, θ_1, that is specified by H_1.

(a) For what value of θ_1 is this evidence for H_1 versus H_0 strongest?

(b) For the value of θ_1 in (a), what is the likelihood ratio, f_1/f_0?

(c) If k represents the number of consecutive draws producing white balls in the urn scheme of section 1.6, to what value of k does the likelihood ratio in (b) correspond?

(d) Discuss these results.

1.5 For the same model and hypotheses as in Exercise 1.4, suppose we choose some number $k > 1$ and interpret observations as strong evidence in favor of H_1 over H_0 when f_1/f_0 exceeds k.

(a) What value of θ_1 maximizes the probability, when H_0 is true, of finding strong evidence in favor of H_1?
(b) What is the maximum probability in (a)?
(c) Compare the bound in (b) with the universal bound, $1/k$, that was derived in section 1.4.
(d) When $\theta = 0$, what is the probability that for some value of $\theta_1 > 0$ the hypothesis $H_1: \theta = \theta_1$ will be better supported than H_0 by a factor of at least k? That is, what is the probability of observing values $x_1, x_2, \ldots, x_n$ for which some positive θ can be found that is better supported than $\theta = 0$?

1.6 Consider testing H_1: $X \sim f_1$ versus H_2: $X \sim f_2$ on the basis of an observation on X, with the goal of minimizing the sum of the two error probabilities, $\alpha + \beta$. Show that the best test procedure is to 'choose H_2 if the observation is evidence supporting H_2 over H_1; otherwise choose H_1'. The critical region that corresponds to this rule is $R_0 = \{x; f_2(x) > f_1(x)\}$. Show also that the critical region $R_1 = \{x; f_2(x) \geq f_1(x)\}$ is just as good as R_0. [*Hint*: $\alpha + \beta = \sum_R f_1(x) + 1 - \sum_R f_2(x)$.]

1.7 For the bivariate normal probability model in section 1.13, show that when $k = \exp\{(1.96)^2/2\} = 6.82$ the $1/k$ likelihood interval (1.8) is a 95% confidence interval for y. That is, show that the **random interval** defined by replacing x in (1.8) by a random variable which has the conditional probability distribution of X, given $Y = y$, contains the point y with probability 0.95.

CHAPTER 2

Neyman–Pearson theory

2.1 Introduction

Much of contemporary statistical practice consists of using the methods of hypothesis testing, estimation, and confidence intervals in order to represent and interpret the evidence in a given set of observations. These same methods are used for other purposes as well, but here we are concerned only with their role in interpreting observed data as evidence, as typified by their conventional use in research reports in scientific journals. In particular, we are concerned with the rationale behind such applications. The most widely taught statistical theory, which is based on a paradigm of Neyman and Pearson (1933), explicitly views these statistical methods as solutions to problems of a different kind, so that these evidential applications fall outside the scope of that theory. In this chapter we describe the Neyman–Pearson theory and look at problems that arise when its results are used for interpreting data as evidence.

2.2 Neyman–Pearson statistical theory

At the heart of Neyman–Pearson theory is the problem of testing two simple hypotheses, which was considered briefly in section 1.8. In Chapter 1 we examined a rule for interpreting observations $X = x$ as evidence for one hypothesis *vis-à-vis* another. Neyman–Pearson theory is not concerned with such interpretations; instead, its focus is on using the observations to make a choice between the two hypotheses. In the words of Neyman (1950, p. 258):

> The problem of testing a statistical hypothesis occurs when circumstances force us to make a choice between two courses of action: either take step A or take step B...

He goes on to explain that he is considering situations when the desirability of actions A and B depend on the unknown probability distribution of a random variable X, and our action is to be determined by the observed value of X. Action A is preferred if the distribution belongs to one set of possible distributions for X and B is preferred if it belongs to another set:

> *any rule R prescribing that we take action A when the sample point ... falls within a specified category of points, and that we take action B in all other cases, is a test of a statistical hypothesis.*
>
> (Neyman, 1950, p. 258)

He then lets H denote the set of distributions where action A is preferred, and $\bar{H}$ the set where B is preferred.

> The choice between the two actions A and B is interpreted as the *adoption* or the *acceptance* of one of the hypotheses H or $\bar{H}$ and the *rejection* of the other. Thus, if the application of an adopted rule... leads to action A, we say that the *hypothesis H is accepted* (and, therefore $\bar{H}$ is rejected). On the other hand, if the application of the rule leads to action B, we say that the *hypothesis H is rejected* (and, therefore, the hypothesis $\bar{H}$ is accepted). Frequently it is convenient to concentrate our attention on a particular one of the two hypotheses H and $\bar{H}$. To do so, one of them is called the *hypothesis tested.* The outcome of the test is then reduced to either accepting or rejecting the hypothesis tested. Plainly it is immaterial which of the two alternatives H and $\bar{H}$ is labelled the hypothesis tested. (Neyman, 1950, p. 259)

Neyman then warns against interpreting the result of a test to mean anything except a decision to choose one action or the other:

> The terms 'accepting' and 'rejecting' a statistical hypothesis are very convenient and are well established. It is important, however, to keep their exact meaning in mind and to discard various additional implications which may be suggested by intuition. Thus, to accept a hypothesis H means only to decide to take action A rather than action B. This does not mean that we necessarily believe that the hypothesis H is true. Also if the application... 'rejects' H, this means only that the rule prescribes action B and does not imply that we believe that H is false. (Neyman, 1950, p. 259)

Here Neyman places his theory squarely in the domain of the second of the physician's three questions of Chapter 1, 'What should I *do*?'. He is careful to deny explicitly that it is intended to answer the first question, 'What do I believe?', while ignoring altogether the third question, the one that we are concerned with, 'How should I interpret this observation as evidence?'.

How are these decision rules to be evaluated? What criteria determine whether one test is better than another? The view of the Neyman–Pearson school is that a statistical test procedure should be evaluated in terms of its error probabilities, i.e. the probability of rejecting H when it is true, and the probability of accepting H when $\bar{H}$ is true. A good test is one with small error probabilities. In the simple-versus-simple case these are just the Type I and Type II error probabilities, α and β. If two tests have the same α, then the one with the smaller β is the better test. The fundamental lemma of Neyman and Pearson (1933) shows how, for any fixed value of α, to find the best test, the one with smallest β. If we use such a test then we still risk making a Type I error, but we have controlled that risk at α. And we can be sure that any test with a smaller Type II risk than ours carries a larger Type I risk.

The Neyman–Pearson theory of hypothesis testing, with its attractive pragmatic focus on minimizing the probabilities of making errors, provides a powerful paradigm that dominates contemporary statistical theory. Wald (1939; 1950) made basic generalizations showing that much of the rest of statistics could be modelled after the optimal decision-making approach of the Neyman–Pearson theory of hypothesis testing. For this reason the general theory is sometimes called the Neyman–Pearson–Wald theory (e.g. Basu, 1975; Carnap, 1950; Efron, 1986), and it views the basic subject matter of statistics as a collection of decision-making problems that are analogous to the hypothesis-testing problem in that they are formulated in terms of choosing between alternative actions. In the hypothesis-testing problem there are only two actions, corresponding to the two hypotheses. In estimation, the actions correspond to values of the parameter being estimated; the goal is to choose a value that is close to the true parameter. And in the confidence interval problem the actions correspond to sets of parameter values, the goal being to choose a set that contains the true value.

So according to the Neyman–Pearson–Wald formulation, statistics is primarily concerned with using observations to choose from a specified set of actions, the desirability of the actions being dependent on which probability distribution is generating the observations. Neyman's expression for this process is **inductive behavior**: 'If a rule R unambiguously prescribes the selection of action for each possible outcome . . . , then it is a rule of inductive behavior' (Neyman, 1950, p. 10). In his view, the generalized Neyman–Pearson theory encompasses the whole of statistics:

> **Scope of Mathematical Statistics.** *Mathematical statistics is a branch of the theory of probability. It deals with problems relating to performance characteristics of rules of inductive behavior based on random experiments.* (Neyman, 1950, p. 11)

This extravagant view of the scope of Neyman–Pearson theory has been widely accepted:

> In recent years, Statistics has been formulated as the science of decision making under uncertainty. This formulation represents the culmination of many years of development and, for the first time, furnishes a simple and straightforward method of exhibiting the fundamental aspects of a statistical problem. (Chernoff and Moses, 1959, p. vii)

And it remains fundamental – a recent course announcement for a basic statistical theory course at my own university (Johns Hopkins) explained that 'Statistics is the science of using data to make decisions'.

Neyman–Pearson theory formulates a statistical problem in terms of choosing from among a specified set of actions. A solution is a *procedure for choosing* an action (a 'rule of inductive behavior'), a protocol that specifies for every possible value of the random variable X whose probability distribution is under study, what action is to be taken if that value is observed. A solution to a testing problem may take the form 'Choose H_2 if $\bar{X} \geq 7$; otherwise choose H_1'. A solution to an estimation problem might be 'Estimate θ by $\bar{X}$' or '... by $\sum(X_i - \bar{X})^2/n$'.

The basic tenet of Neyman–Pearson theory is that solutions to statistical problems, that is, statistical procedures, should be evaluated in terms of their probabilistic properties ('performance characteristics', in Neyman's words). These properties measure the expected, or long-run average, performance of the procedures – a procedure with good probabilistic properties will, if used repeatedly, give good performance, on average. In the simple-versus-simple hypothesis-testing problem, procedures are evaluated in terms of their Type I and Type II error probabilities. An estimation procedure, or estimator, associates with every possible observation x an estimate $t(x)$ of the unknown parameter. If θ denotes this parameter and $X = x$ is observed, then $t(x)$ is used as an estimate of θ. The probabilistic properties that are most popular for evaluating estimators are the expected error, or bias, $\mathrm{E}[t(X) - \theta]$, the variance, $\mathrm{var}[t(X)]$, and the expected squared error, $\mathrm{E}[t(X) - \theta]^2$. A confidence interval procedure associates with every x an interval, $(\ell(x), u(x))$ of parameter values, and two key properties are the probability that the interval will contain the true value of the parameter, $\Pr(\ell(X) < \theta < u(X))$, and the expected width of the interval, $\mathrm{E}[u(X) - \ell(X)]$.

To illustrate the estimation theory we can again consider repeated independent draws from an urn. If X is the number of white balls in ten draws, then to estimate the proportion of white balls in the urn, θ, we might first consider the estimator $t_1(X) = X/10$, which estimates θ by the proportion of draws that produce white balls. This estimator is unbiased, $\mathrm{E}[t_1(X) - \theta] = 0$, and its variance and expected squared error are both equal to $\theta(1-\theta)/10$. An alternative estimator is $t_2(X) = \frac{1}{2}$, which simply ignores X and estimates θ to equal $\frac{1}{2}$ regardless of the value of X. This estimator has a bias, of course, $\mathrm{E}[t_2(X) - \theta] = \frac{1}{2} - \theta$. But its variance is small, $\mathrm{var}[t_2(X)] = 0$, and its expected squared error is $(\frac{1}{2} - \theta)^2$. A third competitor is $t_3(X) = (1-w)t_1(X) + \frac{1}{2}w$, where $w = 1/(1 + 10^{1/2})$ or about 0.24. This estimator represents a compromise between t_1 and t_2. Its bias is $\mathrm{E}[t_3(X) - \theta] = w(\frac{1}{2} - \theta)$, its variance is $\theta(1-\theta)w^2$, and its expected squared error is simply $(\frac{1}{2}w)^2$.

In terms of bias, t_1 is the best of the three estimators; in terms of the variance t_2 is best; and in terms of the expected squared error, Figure 2.1 shows that t_1 is best if θ is less than 0.17 or greater than 0.83, t_2 is best if $0.38 < \theta < 0.62$ (but much the worst if θ is close to zero or one), while t_3 is best for the remaining values of θ, $0.17 < \theta < 0.38$ and $0.62 < \theta < 0.83$.

This situation is typical – there is no best procedure. One is best with respect to one performance measure, but for a different criterion another procedure is best, while for a third criterion, one procedure is better for some values of the unknown parameter and another is better for other values. Here $t_1(X)$ happens to be the best,

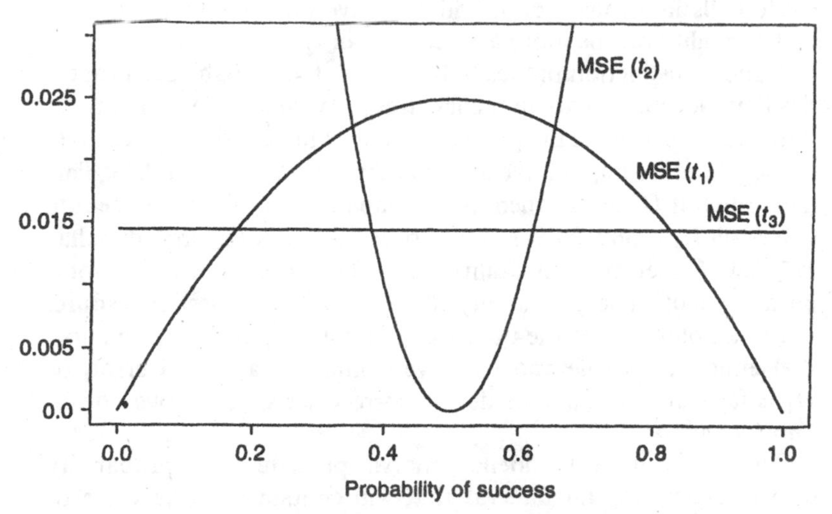

Figure 2.1 *Mean square errors for three estimators of a binomial probability.*

in one sense, among all of the estimators that are unbiased – no other unbiased estimator has smaller variance at any value of θ. How good $t_1(X)$ is, as measured by its variance or its expected squared error, $\theta(1-\theta)/10$, depends on the unknown value of θ. By contrast, the expected squared error of $t_3(X)$ is $(\frac{1}{2}w)^2$, the same for all θ (Figure 2.1). Now $t_3(X)$ is the best of all possible estimators with respect to an important property, the maximum possible value of the expected squared error: the use of $t_3(X)$ ensures that the expected squared error does not exceed $(\frac{1}{2}w)^2$, regardless of the true value of θ. Every other estimator has some values of θ at which the expected squared error exceeds this bound. The estimator $t_2(X) = \frac{1}{2}$ is the best possible estimator if it happens that $\theta = \frac{1}{2}$, and it is much better (smaller expected squared error) than either of the other two if θ is very close to $\frac{1}{2}$.

The user must decide whether the properties of $t_1(X)$, $t_2(X)$, or $t_3(X)$ are more important for his or her particular problem. Although a procedure may be disfavored because it lacks a property that is judged to be important (such as unbiasedness or minimizing the maximum possible mean squared error), neither procedure can be described as 'incorrect' or 'invalid'. For estimating the probability of 'heads' for a coin in my pocket, on the basis of ten tosses, I would use $t_2(X) = \frac{1}{2}$ in preference to t_1 or t_3, because I know that the actual probability is very close to one-half, certainly between 0.4 and 0.6, and in this range t_2 is much the best estimator of the three. On the other hand, for estimating the proportion of white balls in an urn about which I have no prior knowledge, t_1 and t_3 might both be more attractive than t_2.

A subtle distinction proves to be critical: the probabilistic properties that Neyman–Pearson theory uses to evaluate a decision procedure are properties of the procedure, not of its results. For example, a test procedure might have a Type I error probability of 0.05. This means that if H_1 is true then the probability that this test procedure will reject H_1 is only 0.05. It does not mean that if the procedure has rejected H_1 then the probability that a Type I error has been committed is 0.05. The probability, 0.05, refers to the test procedure, not to an outcome of the procedure. Having rejected H_1, we know that either H_1 is true and we have committed a Type I error, or H_1 is false and we have made the correct decision; but we do not know which.

Similarly, a 95% confidence interval procedure has probability 0.95 of generating an interval that will contain the true value of the target parameter, say θ. For example, if X_i, $i = 1, 2, \ldots, n$, are independent and identically distributed (i.i.d.) $N(\theta, 1)$ then $(\bar{X} - 1.96/\sqrt{n}, \bar{X} + 1.96/\sqrt{n})$ defines a 95% confidence interval procedure. The interval is random, depending on the random variable $\bar{X}$; it will contain θ if and only if $\bar{X}$ falls in the interval $(\theta - 1.96/\sqrt{n}, \theta + 1.96/\sqrt{n})$, and since $\bar{X}$ has a $N(\theta, 1/n)$ distribution the probability that this will occur is 0.95. But after $\bar{X}$ has been observed to equal $\bar{x}$ and we have the specific interval, $(\bar{x} - 1.96/\sqrt{n}, \bar{x} + 1.96/\sqrt{n})$, the probability statement no longer applies – either θ is in this interval or it is not, and we do not know which. Both the interval and the parameter θ are fixed, not random. We can say that we used a procedure that generates an interval containing θ 95% of the time, and that in this instance it generated the interval $(\bar{x} - 1.96/\sqrt{n}, \bar{x} + 1.96/\sqrt{n})$. But we cannot say that the probability that θ is in *this* interval is 0.95. That this is not semantic hair-splitting is illustrated by the confidence interval example in the next section, while Exercise 2.3 gives an example of a 95% confidence interval that contains all the possible values of θ!

2.3 Evidential interpretation of the results of Neyman–Pearson decision procedures

Contrary to the views quoted in the preceding section, many statistical applications in research are not well represented by the Neyman–Pearson model of choosing between alternative actions. And in addressing these applications many statisticians explicitly reject that formulation, instead describing the problems in terms of 'inductive reasoning' (Fisher, 1959, p. 109), representing 'what the data say' (Cox, 1958, p. 359), finding an 'index to or measure of weight of evidence' (Cornfield, 1966, p. 18), 'summarization of evidence' (Cox and Hinkley, 1974, p. 56), etc. Nevertheless, one approach to such applications employs many of the same statistical tools, methods, and even concepts as Neyman–Pearson theory. The same general problem areas are identified – hypothesis testing, estimation, and confidence intervals – and many of the same tests, estimators, and confidence interval procedures are used. Moreover, the procedures are evaluated, just as in Neyman–Pearson theory, in terms of their probabilistic properties – size, power, bias, variance, etc.

In these applications the probabilistic properties not only determine which procedure will be used, as envisioned by the Neyman–Pearson theory; after a procedure has been used its properties are reported alongside the result. Thus in addition to the fact that our test, when applied to the data observed in this experiment, leads us to choose H_2, our published report also describes the test's size and, sometimes, its power. In addition to the value of the estimate, we report that the estimator (i.e. the estimation procedure used to derive this estimate) is unbiased, if it is, and we state the (estimated) value of its standard error. In addition to describing the confidence interval that our observations produced, we report the confidence coefficient, the coverage probability of the procedure that generated this particular interval.

The reason for reporting not only the result but also the probabilistic properties of the procedure leading to that result is that the procedure is actually being used, not to choose an action, but to indicate 'what the data say', that is, to interpret the data as evidence. The probabilistic properties are used to refine and quantify this interpretation. In these applications, when a test procedure leads to the rejection of H_1 it is not really taken to mean that we or anyone else actually decides to act as if H_1 were false. Rather 'reject H_1' is interpreted as a figure of speech meaning that the data in question are evidence against H_1; and the error probability α, or the two probabilities α and β, are interpreted as somehow measuring the strength of that evidence.

Likewise, although an estimated treatment effect in a published report of a clinical trial might represent a decision to act as if the treatment really has that precise effect, it more commonly is interpreted to mean that the observations are evidence that the effect has approximately the size of the estimate, and the standard error of the estimation procedure is supposed to show how strong the evidence is, a large standard error indicating that it is weak.

A confidence interval is also commonly given an evidential interpretation. In fact it is sometimes recommended that not one but a system of confidence intervals be reported, one for each of a series of confidence coefficients, such as 80%, 90%, 95%, and 99%. This system is not interpreted as some (very complicated) action, in the Neyman–Pearson sense. Rather its interpretation is evidential – it 'summarizes what the data tell us about θ, given the model' (Cox and Hinkley, 1974, p. 227).

Such applications and interpretations, although formally outside the scope of Neyman–Pearson theory, are not entirely unauthorized; one of the leading expositors of the Neyman–Pearson school acknowledges the use of confidence intervals for 'indicating what information is available concerning the unknown parameter' (Lehmann, 1959, p. 4). And Neyman (1976, p. 749) himself wrote that when the two possible results of a hypothesis test are described with reference to a single hypothesis H, namely (1) reject H and (2) do not reject H, 'My own preferred substitute for "do not reject H" is "no evidence against H is found"'.

For these evidential applications the distinction between a statistical procedure and a result of using that procedure is critical. Column 1 in Table 2.1 lists some of the basic procedures pertaining to an unknown parameter θ. They are defined in terms of a generic random variable X and are themselves random. Column 2 lists some of their important probabilistic properties. Column 3 represents the time when a realized value x of the random variable X is observed, and column 4 shows the result that is generated by the statistical procedure when X is observed to take the value x. Finally, column 5 contains examples of statements in which the properties of the procedures are used for interpreting the observation x as evidence about θ.

The statements in the second column concern well-defined probabilistic properties of random variables and are subject to rigorous mathematical verification. The meanings of the statements in the fifth column are less clear. Yet these statements, and others like them, are an integral part of today's dominant statistical methodology, which uses tests, estimates, and confidence intervals for interpreting and representing statistical observations as evidence. Here is how Pratt (1961) described one evidential interpretation of confidence intervals:

> What has made the confidence interval popular is 'indicating what information is available'. Decision problems seem beside this point; a confidence interval probably contains the parameter, and the confidence coefficient measures how probably. But does it? By the formal definition, it no longer does, once we insert numerical values for the endpoints. Then no probability (except 0 or 1) can be attached to the event that the interval contains the parameter: either it does or it doesn't. Unfortunately we don't know which. We think, and would like to say, it 'probably' does; we can invent something else to say, but nothing else to think. We can say to an experimenter, 'A method yielding true statements with probability .95, when applied to your experiment, yields the statement that your treatment effect is between 17 and 29, but no conclusion is possible about how probable it is that your treatment effect is between 17 and 29'. The experimenter, who is interested not in the method, but in the treatment and this particular confidence interval, would get cold comfort from that if he believed it.

Thus, although the Neyman–Pearson theory of confidence intervals stops at the fourth column of Table 2.1, typical applications

Table 2.1 *Properties of statistical procedures applied to results of the procedures for evidential interpretation of observations*

Procedure (depends on random variable X)	Property of procedure (determined by probability distribution of X)	Observation (realized value of random variable X)	Result of procedure (fixed by observation)	Evidential interpretation (property used to interpret observation)
Confidence interval $(\ell(X), u(X))$	We can be 0.95 confident that the random interval $(\ell(X), u(X))$ will contain θ. (Confidence coefficient: $\Pr(\ell(X) < \theta < u(X)) = 0.95$)	$X = x$	An interval $(\ell(x), u(x))$	The observation x is evidence that θ is in $(\ell(x), u(x))$. Large confidence coefficient means strong evidence.
Estimator $t(X)$	The expected value of $t(X)$ is θ and its standard error is σ. ($\mathrm{E}[t(X)] = \theta$; $\mathrm{var}[t(X)] = \sigma^2$)	$X = x$	An estimate, $t(x)$	The observation x is evidence that θ is near $t(x)$. The smaller σ, the stronger the evidence.
Hypothesis test $H_{\delta(x)}$	Type I and Type II error probabilities are α and β	$X = x$	A hypothesis, $H_{\delta(x)}$ (H_1 if $\delta(x) = 1$; H_0 if $\delta(x) = 0$)	The observation x is evidence in favor of the hypothesis $H_{\delta(x)}$. Small α and β mean strong evidence.

in scientific investigation and reporting, where the objective is to represent the evidence in a given set of observations, proceed to the fifth column.

Is it valid to use Neyman–Pearson theory in this way, interpreting the procedures of hypothesis testing, estimation, and confidence intervals as techniques for representing 'what the data say'? For instance, if a good procedure for testing H_1 versus H_2 leads to acceptance of H_2, does this mean that the data are evidence supporting H_2 over H_1? If a good estimation procedure leads to a specific estimate, does it mean that the data are evidence that the parameter lies near that value? If a good confidence interval procedure leads to the interval (a, b), does it mean that the data are evidence that the parameter lies between a and b? And do the probabilistic properties of the procedures, the error probabilities, standard errors, confidence coefficients, etc., measure the strength of the evidence? All of these questions have the same simple answer – no.

We have already seen in section 1.8 that the above evidential interpretation of Neyman–Pearson test results is not valid – a good test, one whose error probabilities are both very small, can call for choosing H_1 when the evidence favors H_2 and vice versa.

Randomized tests furnish another example of the problems that appear when we try to interpret Neyman–Pearson test procedures as showing 'what the data say'. Suppose we make five draws from an urn in which either (H_1) half of the balls are white or (H_2) three-fourths are white, replacing the ball after each draw. Let X represent the number of white balls that we observe. If we reject H_1 whenever $X = 5$ we will have the best (most powerful) test of size $\alpha = p_1(X = 5) = 1/32$, and if we reject whenever $X = 4$ or $X = 5$ we will have the best test of size $\alpha = 6/32$. If we want the best test having size $\alpha = 0.05$, we must use a randomized test; it calls for rejecting H_1 whenever $X = 5$ as well as rejecting sometimes, but not always, when $X = 4$. Specifically, when $X = 4$ it rejects 12% of the time. We might carry out such a test as follows: if $X = 5$, reject H_1; if $X = 4$, choose a random number U between 0 and 1, and, if $U \leq 0.12$, rejcct H_1; otherwise accept H_1. Our test has the desired size $\alpha = p_1(X = 5) + 0.12p_1(X = 4) = 0.05$, and the fundamental lemma of Neyman and Pearson assures us that there is not a better one – among all tests that have size 0.05 or less, ours has the smallest possible Type II error probability. In particular, ours has smaller Type II error probability than any non-randomized test with $\alpha \leq 0.05$.

But despite its optimality, most statisticians would be reluctant to use this test in applications where the purpose of the statistical analysis is to indicate what the observations say about H_1 *vis-à-vis* H_2. They properly sense that whatever $X = 4$ means as evidence, it is not affected by whether $U \leq 0.12$ or not, and to let their assessment of the evidence depend on this clearly irrelevant event is inappropriate. The evidence about the unknown proportion of the white balls in the urn is the same when $X = 4$ and the test calls for rejecting H_1 as it is when $X = 4$ and the test calls for accepting.

Straightforward evidential interpretation of Neyman–Pearson confidence intervals is also invalid. This is illustrated in the following example, which is derived from a famous one of Cox (1958, p. 360). It concerns an experiment that is conducted in two stages. At the first stage we simply toss a coin; the outcome of the toss determines what happens at the second stage. If the coin falls heads then we observe a random variable X with a $N(\theta, \sigma^2)$ probability distribution. But if the coin falls tails we observe k independent random variables $X_1, X_2, \ldots, X_k$, all having the same $N(\theta, \sigma^2)$ distribution. That is, at the second stage we make either one or k observations, depending on the result of the coin toss. The value of σ^2 is known and we want a 95% confidence interval for θ with short expected length.

If instead of using a coin toss we choose the sample size, say n, deliberately, then it is well known that $\bar{X} \pm 1.96\sigma/n^{1/2}$ represents the best (shortest expected length) 95% confidence interval procedure. Thus in our two-stage experiment we might consider procedure A: if the coin falls heads and $X = x$ is observed, use the interval $x \pm 1.96\sigma$; if it falls tails and $X_1 = x_1, \ldots, X_k = x_k$ are the observations, use $\bar{x} \pm 1.96\sigma/k^{1/2}$. That is, regardless of which sample size our coin toss tells us to use, we employ the best 95% confidence interval procedure for that sample size. This is certainly a reasonable procedure; but we can do better.

When the sample size is determined by a coin toss the best (shortest expected length) 95% confidence interval depends on the value of k, the number of observations we make if the coin falls tails. To make the example concrete we let $k = 100$. In that case the best 95% confidence interval procedure (B) uses $x \pm 1.68\sigma$ if the coin falls heads and $\bar{x} \pm 2.72\sigma/10$ if it falls tails.

Both A and B are valid 95% confidence interval procedures: if the coin falls heads, the coverage probability of A is 0.95 while that of B is 0.91; if it falls tails then A again covers θ with probability 0.95 while B's probability is 0.99. Thus A's overall coverage probability is $\frac{1}{2} \times 0.95 + \frac{1}{2} \times 0.95 = 0.95$ and B's is the same: $\frac{1}{2} \times 0.91 + \frac{1}{2} \times 0.99 = 0.95$. But the expected width of an interval produced by A is $1.96\sigma + 1.96\sigma/10 = 2.16\sigma$, while that of an interval produced by B is only $1.68\sigma + 2.72\sigma/10 = 1.95\sigma$; B generates intervals that are, on the average, about 10% shorter. Note that even if we increase the value of k in procedure A, its expected interval width cannot be made as small as that produced by B with $k = 100$.

Although B is the better procedure, if we apply it to the observations from an experiment for the purpose of indicating what those observations say about θ, then it appears to be misleading in every instance. When we take only one observation, it seems wrong to present $x \pm 1.68\sigma$ as a 95% confidence interval – the confidence coefficient, 0.95, seems too large. Similarly, when the coin falls tails and we make k observations, it seems wrong to attach to the interval $\bar{x} \pm 2.72\sigma/k^{1/2}$ a confidence coefficient of only 0.95.

Another way to look at this example is to compare the cases when we have taken 100 observations deliberately and when we have made this choice of sample size by the toss of a coin. In the first case the best 95% confidence interval procedure uses $\bar{x} \pm 1.96\sigma/10$, while in the second the best procedure uses $\bar{x} \pm 2.72\sigma/10$. Many people agree that, whatever the evidence concerning θ in the observations $x_1, \ldots, x_{100}$, it is unaffected by whether the number of observations was fixed by considerations of costs, for example, or by a coin toss. That is, for interpreting a given set of observations as evidence about θ, it does not matter whether they arose in the first case or the second. If a particular interval is appropriate for showing what the data say in one case, then it is also appropriate in the other. The Neyman–Pearson theory leads to different results in two situations where the evidence is the same, and in applications where the purpose of the statistical analysis is to represent and interpret the data as evidence, this is unacceptable.

Problems appear also when Neyman–Pearson estimation theory is used in applications where the goal is evidential interpretation. We have seen a simple example of this in section 2.2 – for estimating the probability of heads on the basis of ten tosses of a coin the estimator $t_2(X) = \frac{1}{2}$ is a good one in the Neyman–Pearson sense if performance is measured in terms of expected squared error. Now our sample does constitute evidence concerning the probability of heads, but this estimator, which ignores the sample altogether, in no sense represents that evidence.

Here is a much less trivial example. Suppose that one colleague brings me his measurements on the length of butterfly wings in

Ecuador, another brings her observations on tensile strength of wire samples, and I have some data of my own showing the weight loss in laboratory rats on a special diet. If all of the measurements have normal probability distributions, James and Stein (1961) showed that the naive procedure which uses the three sample means to estimate their respective parameters can be improved. Specifically, the naive procedure is not as good as one devised by Stein that uses a statistic depending on all three means, the butterfly wings, the tensile strengths, and the rat weights, to estimate the mean length of butterfly wings, another statistic depending on all three to estimate the mean tensile strength, etc. The James–Stein estimation procedure is better in the sense that the average of the three expected squared errors is smaller with that procedure than with the naive one, no matter what the true values of the three parameters are. But for interpreting our observations as evidence about butterflies, etc., this estimation procedure makes no sense. Whatever evidence we have about butterfly wings is contained in the butterfly data. (The likelihood ratio measuring the relative support for two values of the wing parameter depends only on the butterfly data.) It is inappropriate to allow our assessment of that evidence to depend on irrelevant observations related to wires and rats. Just as in the Cox confidence interval example, we find that the procedure that is better in the Neyman–Pearson sense of expected or average performance is unsatisfactory as a tool for interpreting data as evidence, because it leads to different results in situations where the evidence is the same.

Of course, not all attempts to use Neyman–Pearson methodology for evidential interpretation of data produce results that are as obviously unsatisfactory as the examples above. If they did, the discipline of statistics would look very different than it does today. In countless applications every day, statistical evidence is interpreted, analyzed, and reported in terms of hypothesis tests, estimates, and confidence intervals. Many of these applications seem to be reasonable and helpful, both to experimenters and to readers of their research reports, for representing and communicating 'what the data say'. We will pursue this point in Chapters 3 and 4.

Although Neyman–Pearson test procedures do not have a valid evidential interpretation in general, there is one interesting exception. In that special case the interpretation is derived from the law of likelihood. Suppose it is reported that a test with small error probabilities, α and β, has led to the choice of H_2. In this situation it seems reasonable to claim that the report represents evidence favoring H_2 over H_1, and that the smaller α and β are, the stronger the evidence is. The test procedure usually chooses the correct hypothesis (because α and β are small), and in this instance it has chosen H_2. Is this not evidence that H_2 is correct? Is it not right to presume that what usually happens (i.e. the procedure chooses correctly) *has* happened, in the absence of any evidence to the contrary? The law of likelihood confirms this judgement. The key here is that the evidence we are evaluating is not the observation, $X = x$, that led to the choice of H_2, but simply an indicator showing which hypothesis was chosen – instead of observing X itself, we see only $Z(X)$, where $Z(x) = 2$ if x is in the critical region (so H_2 is chosen) and $Z(x) = 1$ otherwise (and H_1 is chosen). Thus when we observe $Z = 2$ we have the likelihood ratio of $\Pr(Z = 2|H_2)/\Pr(Z = 2|H_1) = (1 - \beta)/\alpha$ in favor of H_2 over H_1. That is, according to the law of likelihood, the report 'A test of H_1 versus H_2 having size α and power $1 - \beta$ led to rejection of H_1 in favor of H_2' represents evidence favoring H_2 by the factor $(1 - \beta)/\alpha$ (Barnard, Jenkins, and Winsten, 1962, p. 331; see also Birnbaum, 1977). Note that precisely the same reasoning applies when we are told that the test has led to the choice of H_1, rather than H_2. The report 'A test of H_1 versus H_2 having size α and power $1 - \beta$ led to acceptance of H_1' represents evidence favoring H_1 by the factor $(1 - \alpha)/\beta$.

That $Z = 2$ is evidence for H_2 over H_1 does not mean that the observation, $X = x$, on which the test is based is evidence for H_2 over H_1. A data reduction has been made, and evidence has been discarded. That is, although we can give a valid evidential interpretation to the result of a Neyman–Pearson test procedure, that interpretation does not necessarily represent even crudely the evidence in the original observation $X = x$. In section 1.7 we observed X itself and found that some of the values that fell in the critical region (leading to rejection of H_1 for H_2) were in fact evidence favoring H_1 over H_2. Here we observe only whether X is in the critical region or not. Our conclusion that when H_2 is selected we have evidence in favor of H_2 over H_1 refers to the evidence in the limited information given to us, not to the evidence in the observation $X = x$, that caused H_2 to be selected. Thus, if we are not told the value of x, but only that it produced the test result $Z = 2$, say, then we can give a proper evidential interpretation of this very limited information (via the likelihood ratio $(1 - \beta)/\alpha$). But it is not a proper evidential interpretation of x.

The hypothesis-testing procedures that are most often used for interpreting and reporting scientific data are not of the

Neyman–Pearson variety. Before turning to the more commonly used test procedures, which are discussed in Chapter 3, we consider a place in science where Neyman–Pearson theory does play an important role.

2.4 Neyman–Pearson hypothesis testing in planning experiments: choosing the sample size

We have observed that Neyman–Pearson tests are not designed for interpreting statistical evidence, and that their use for that purpose can lead to serious errors in which observations that are evidence supporting H_1 over H_2 are given the opposite interpretation. Strict Neyman–Pearson test procedures are in fact rarely used for interpreting and reporting scientific data, but they are routinely used in another important phase of research. When a study or experiment is being planned, the researcher often uses Neyman–Pearson theory to determine how many observations will be made. He models the study as a procedure for choosing between two hypotheses, H_1 and H_2, and specifies the maximum tolerable error probabilities, α and β. Then two objectives, stated in terms of the Neyman–Pearson hypothesis-testing paradigm, determine the sample size: 'We want to be pretty sure (probability $1-\alpha$ or greater) that we will not reject H_1 when it is true, and also pretty sure (probability $1-\beta$ or greater) that we will reject H_1 when H_2 is true'.

For any sample size we can choose to test at any size we like, so we can always accomplish the first objective. But in order to accomplish the second we must make the sample size large enough.

This approach leads to standard formulas for the number of observations required. For example, consider the simple case of independent $N(\theta, \sigma^2)$ observations, where the variance σ^2 is known from pilot data or from results of previous studies, with hypotheses H_1: $\theta = \theta_1$ and H_2: $\theta = \theta_1 + \delta$. The usual calculation shows that the number of observations must be at least

$$n_{NP} = [(z_{1-\alpha} + z_{1-\beta})\sigma/\delta]^2, \tag{2.1}$$

where $z_{1-\alpha}$ is the $100(1-\alpha)$th percentile of the standard normal distribution. Using a sample size $n \geq n_{NP}$ ensures that a test with size α will have power of at least $1-\beta$: the Type I and Type II error probabilities will not exceed the specified values, α and β.

This approach to determining sample size is often used in studies whose actual purpose is more accurately described in terms of evidence than decisions. The objective is not really to choose between $\theta = \theta_1$ and $\theta = \theta_1 + \delta$, but to generate evidence about θ, with particular interest in how strongly that evidence supports one of these two values, θ_1 and $\theta_1 + \delta$, versus the other. When this is true, the Neyman–Pearson approach is unsatisfactory. We will show that in the simple important case of the normal probability distribution described above, the sample size n_{NP} is too small to ensure that the researcher has an adequate chance to meet his actual objectives, and that using the Neyman–Pearson hypothesis-testing procedure to interpret the data as evidence leads to misinterpretation with high frequency. That is, the misinterpretations that were shown to be possible in section 2.3 are not confined to extreme or pathological cases, but are common when the sample size is determined by the 'usual' formula. Specifically, we will see that when $\alpha = 0.05$ and $\beta = 0.20$ the evidence will be misinterpreted more than 30% of the time.

The objectives can be restated as 'We want to be pretty sure (probability $1 - \alpha$ or greater) that we will not find strong evidence in favor of H_2 when H_1 is true, and also pretty sure (probability $1 - \beta$ or greater) that we will find strong evidence in favor of H_2 when H_2 is true'. Suppose that these are our actual objectives, but that we use the Neyman–Pearson paradigm, determining our sample size by equation (2.1).

First, we ask 'How often will the study produce strong evidence?'. The results, $X = x$, will be strong evidence for H_2 over H_1 if (and only if) the likelihood ratio L_2/L_1 is at least k, where k is determined by the expression 'strong' evidence. Now most readers will agree that, in the canonical urn scheme of section 1.6, the evidence in favor of the 'all white' urn over the 'half white' one, when the number of consecutive white balls observed is two (a likelihood ratio of $2^2 = 4$) is only weak, but that six consecutive white balls (a likelihood ratio of $2^6 = 64$) are not just 'strong' but 'quite strong' evidence. Thus we will focus on values of $k = 8$, 16, 32, corresponding to 3, 4, or 5 white balls in the urn scheme.

The likelihood ratio for $H_2\colon \theta = \theta_1 + \delta$ versus $H_1\colon \theta = \theta_1$, where θ is the mean of the normal distribution, equals

$$\exp\{[\bar{x} - (\theta_1 + \delta/2)]n\delta/\sigma^2\},$$

so that we have strong evidence for H_2 when we have observations x for which this quantity exceeds k, that is, for which

$$n^{1/2}(\bar{x} - \theta_1)/\sigma > n^{1/2}\delta/2\sigma + \sigma\ln(k)/\delta n^{1/2}.$$

Table 2.2 *Probabilities of undesirable results when n is chosen so that* $\alpha = 0.05$ *for* $\beta = 0.20, 0.05$*: (a) finding strong evidence in favor of false hypothesis; (b) failing to find strong evidence in favor of true hypothesis*

	(a) $\Pr_1(L_2/L_1 \geqslant k)$		(b) $\Pr_2(L_2/L_1 < k)$	
k	$\beta = 0.20$	$\beta = 0.05$	$\beta = 0.20$	$\beta = 0.05$
8	0.019	0.011	0.342	0.156
16	0.009	0.006	0.449	0.212
32	0.004	0.003	0.560	0.277

If the sample size is the value n_{NP} given in expression (2.1), then the right-hand side of this inequality equals $c/2 + \ln(k)/c$, where $c = z_{1-\alpha} + z_{1-\beta}$, and the probability of finding strong evidence in favor of H_2 when H_1 is true is

$$\Pr_1(L_2/L_1 \geq k) = 1 - \Phi(c/2 + \ln(k)/c). \tag{2.2}$$

It is easily shown that this probability of misleading strong evidence is the same as the probability of misleading evidence in the other direction, $\Pr_2(L_1/L_2 \geqslant k)$. Similarly, the probability of finding strong evidence for H_2 when H_2 is true is the same as the probability of finding strong evidence for H_1 when H_1 is true, and that probability is

$$\Pr_2(L_2/L_1 \geq k) = 1 - \Phi(\ln(k)/c - c/2). \tag{2.3}$$

Table 2.2 gives the values of expressions (2.2) and (2.3) for selected values of k when $\alpha = 0.05$ and $\beta = 0.20$ and 0.05.

If the researcher actually wants to be pretty sure (probability at least 0.95) that the study will not produce strong evidence supporting H_2 when H_1 is true, then columns 1 and 2 show that for both $\beta = 0.20$ and $\beta = 0.05$ the sample size n_{NP} is adequate. But so are smaller ones. In fact it is easy to show (Exercise 1.5) that the probability of misleading evidence, $\Pr_1(L_2/L_1 \geq k)$, cannot exceed $1 - \Phi(\sqrt{2\ln(k)})$ for any choice of n, θ_1, and δ. For $k = 8$, 16, 32 this bound equals 0.021, 0.009, and 0.004, respectively. That is, the choice of a reasonable value of k ensures that the probability of generating misleading evidence is small, less than 0.021 when $k = 8$ and 0.01 when $k = 16$, regardless of n.

But columns 3 and 4 show that the sample size n_{NP} is not adequate with respect to producing strong evidence in favor of H_2 when H_2 is true. When β is set at 0.20 and H_2 is true, a sample of size n_{NP} will fail to produce strong evidence for H_2 more than one-third of the time (column 3). That is, the sample is not large enough that the researcher can be pretty sure that, if H_2 is true, the study will produce strong evidence in its favor. And column 4 shows that when $\alpha = \beta = 0.05$, the probability of failing to produce strong evidence in favor of H_2 when it is true is greater than 0.15, three times the value at which the researcher probably thought he was controlling this risk when he fixed the Type II error probability at $\beta = 0.05$.

If the study is structured as a Neyman–Pearson testing procedure, it always leads to a decision, to choose H_1 or to choose H_2. It does not always lead to strong evidence; in fact, Table 2.2 shows that at $\alpha = 0.05$, when either of the two hypotheses is true the study will fail to produce evidence strong enough to give a likelihood ratio as large as 8 in favor of one or the other about one-third of the time $(0.342 - 0.019 = 0.323)$ when $\beta = 0.20$, and about 15% of the time when $\beta = 0.05$.

In this example, if either hypothesis is true the probability of producing strong evidence supporting that hypothesis over the other one is the same, $\Pr_2(L_2/L_1 \geq k) = \Pr_1(L_1/L_2 \geq k)$. To ensure that this probability is at least 0.95, we need at least $n_L = \{1.645 + [(1.645)^2 + 2\ln(k)]^{1/2}\}^2\sigma^2/\delta^2$ observations (Exercise 2.1). Table 2.3 shows these values for $k = 8$, 16, 32. They are larger than the value n_{NP} given by the Neyman–Pearson formula (2.1) with $\alpha = \beta = 0.05$, which equals 10.824 times σ^2/δ^2. For instance, to be pretty sure (probability at least 0.95) that we will obtain strong evidence in favor of the true hypothesis ($LR \geqslant 8$) we require about two-thirds again as many observations as are required to achieve Type I and Type II error probabilities of $\alpha = \beta = 0.05$: $n_L/n_{NP} = 18.191/10.824 = 1.68$. Even if we reduce α to 0.025, so that n_{NP} is the sample size required for a two-sided Neyman–Pearson test with Type I error rate $\alpha = 0.05$, formula (2.1) gives $n_{NP} = 12.996\sigma^2/\delta^2$, so that $n_L/n_{NP} = 1.40$. We actually need 40% more observations.

Perhaps we have chosen the value of k that identifies 'pretty strong evidence' badly. Perhaps $k = 8$ is more extreme than we realize, and a more enlightened choice of this critical value, say $k = 4$, would produce a sample size n_L that agrees with the Neyman–Pearson value n_{NP}. Is there some value of $k > 1$ for which $n_L = n_{NP}$? No. Exercise 2.1 shows that for any specified k we can make $\Pr_1(L_1/L_2 \geq k) = \Pr_2(L_2/L_1 > k) = 0.95$ by choosing n large enough, $n \geq n_L(k)$, where the required sample size increases as k increases. It also shows that for $k = 1$ the required sample size,

Table 2.3 *Sample size n_L required to give the stated probability of producing strong evidence for the true hypothesis, and (α, β) values for which formula (2.1) gives the required sample size*

	Probability = 0.95			Probability = 0.80		
		$\alpha = 0.05$	$\alpha = \beta$		$\alpha = 0.05$	$\alpha = \beta$
k	$n_L\delta^2/\sigma^2$	α β	α β	$n_L\delta^2/\sigma^2$	α β	α β
8	18.191	(0.05, 0.004)	(0.016, 0.016)	9.292	(0.05, 0.080)	(0.064, 0.064)
16	20.408	(0.05, 0.002)	(0.012, 0.012)	11.175	(0.05, 0.045)	(0.047, 0.047)
32	22.557	(0.05, 0.001)	(0.009, 0.009)	13.004	(0.05, 0.025)	(0.036, 0.036)

$n_L(1)$, equals the sample size n_{NP} given by the Neyman–Pearson formula with $\alpha = \beta = 0.05$. This sample size is not adequate for any $k > 1$. We are not finding that the required sample size n_L is greater than the Neyman–Pearson value n_{NP} because we are unwittingly setting our standard too high; n_{NP} is really too small.

If we are willing to settle for a probability as low as 0.80 that the study will produce strong evidence in favor of the true hypothesis, column 4 of Table 2.3 shows that we need only $9.292\sigma^2/\delta^2$ observations, about half the number required for a probability of 0.95. This is still 50% more than the sample size given by the Neyman–Pearson formula (2.1) with $\alpha = 0.05$, $\beta = 0.20$, which is $6.185\sigma^2/\delta^2$.

In order to obtain the required sample size, $n_L(k)$, from the Neyman–Pearson formula, that is, to make $n_{NP} = n_L$, we must choose smaller values of α and β than the conventional ones. Column 2 of Table 2.3 shows the β values needed if $\alpha = 0.05$, and column 3 shows the common value needed if we set $\alpha = \beta$.

Now suppose that we have done the study, taking the number of observations n_{NP} given by formula (2.1). Furthermore, suppose that we use the Neyman–Pearson test procedure to interpret the evidence in our sample. We perform the test of H_1 versus H_2 at level α; because of the way we chose n, we know the test has the specified power, $1 - \beta$. When the test calls for choosing H_2, we will interpret this to mean that the sample represents pretty strong evidence supporting H_2 over H_1 (and vice versa).

We saw in section 2.3 that this interpretation can be wrong, that it can lead to classifying observations as evidence supporting H_2 over H_1 (or vice versa) when the opposite is true. But maybe this is not often the case. Maybe the Neyman–Pearson procedure usually produces a correct interpretation of the evidence in situations like the present example, where the sample size is chosen to control α and β at conventional levels. Let us see.

We will interpret the observations as evidence for H_2 when the test rejects H_1, that is, when $n_1^{1/2}(\bar{x} - \theta)/\sigma > z_{1-\alpha}$, and as evidence for H_1 when this inequality is reversed. When H_1 is true, how often will this interpretation be incorrect? From formula (2.2) we find that

$$\begin{aligned} &\Pr_1(L_2/L_1 < k \,|\, n^{1/2}(\bar{X} - \theta_1)/\sigma > z_{1-\alpha}) \\ &\quad = 1 - [1 - \Phi(c/2 + \ln(k)/c)]/\alpha \qquad (2.4) \end{aligned}$$

(assuming that $c/2 + \ln(k)/c$ is greater than $z_{1-\alpha}$). This is the probability, given that H_1 is rejected in favor of H_2, that the evidence supporting H_2 over H_1 is not strong. Similarly, the probability, given that H_1 is accepted, that the evidence in favor of that hypothesis is not strong is given by

$$\Pr_1(L_1/L_2 < k \mid n^{1/2}(\bar{X} - \theta_1)/\sigma < z_{1-\alpha}) \\ = 1 - [\Phi(c/2 - \ln(k)/c)/(1-\alpha)]. \qquad (2.5)$$

These and the corresponding probabilities of misinterpretation when H_2 is true are given in Table 2.4 for $\alpha = 0.05$ when $\beta = 0.20$ and when $\beta = 0.05$. There we see that if a likelihood ratio of at least $k = 8$ defines 'strong' evidence, then almost two-thirds (0.625) of the 'Type I error' rate results from rejecting H_1 when the evidence in favor of H_2 is not strong.

Overall, if we think that we have strong evidence in favor of the hypothesis that is selected by the testing procedure with $\alpha = 0.05$ and $\beta = 0.20$, Table 2.4 shows that we will be wrong about one-third of the time: when H_1 is true we will accept H_2 5% of the time, and Table 2.4 shows that when that happens, the probability that we will actually have only weak evidence (likelihood ratio less than $k = 8$) in favor of H_2 is 0.625; similarly, we will accept H_1 95% of the time, but when that happens, we will actually have only weak evidence in favor of H_1 30.7% of the time. Thus when H_1 is true we will misinterpret the evidence a fraction $0.05 \times 0.625 + 0.95 \times 0.307 = 0.03 + 0.29 = 0.32$ (about one-third) of the time, usually in the direction of thinking that we have strong evidence for H_1 when we do not. When H_2 is true the same analysis shows that we will misinterpret the evidence about the same fraction of the time, $0.80 \times 0.177 + 0.20 \times 0.906 = 0.14 + 0.18 = 0.32$, incorrectly thinking that we have strong evidence in favor of H_2 (14% of the time) almost as often as we incorrectly think that we have strong evidence for H_1. If we change β so that $\beta = \alpha = 0.05$ then Table 2.4 shows that, regardless of which hypothesis is true, we will misinterpret the evidence 14.4% of the time: $0.05 \times 0.772 + 0.95 \times 0.111 = 0.144$.

At the beginning of this section, we noted that strict Neyman–Pearson procedures are rarely used for interpreting and reporting scientific data as evidence. Table 2.4 shows that this is appropriate – Neyman–Pearson procedures should not be used for that purpose. The more commonly used procedures are discussed in the next chapter. If they are valid, they must rest on a different theoretical basis than the one provided by Neyman and Pearson.

Table 2.4 *Probability that the evidential interpretation of the Neyman–Pearson test result will be incorrect when* $\alpha = 0.05$ *and* $\beta = 0.20$ $(\beta = 0.05)$

k	$\Pr_1(L_2/L_1 < k \mid \text{rej. } H_1)$	$\Pr_1(L_1/L_2 < k \mid \text{acc. } H_1)$	$\Pr_2(L_2/L_1 < k \mid \text{rej. } H_1)$	$\Pr_2(L_1/L_2 < k \mid \text{acc. } H_1)$
8	0.625 (0.772)	0.307 (0.111)	0.177 (0.111)	0.906 (0.772)
16	0.816 (0.871)	0.420 (0.170)	0.311 (0.170)	0.954 (0.871)
32	0.916 (0.930)	0.536 (0.239)	0.450 (0.239)	0.979 (0.930)

2.5 Summary

Neyman–Pearson statistical theory is aimed at finding good rules for choosing from a specified set of possible actions. It does not address the problem of representing and interpreting statistical evidence, and the decision rules derived from Neyman–Pearson theory are not appropriate tools for interpreting data as evidence.

Exercises

2.1 Suppose $X_1, \ldots, X_n$ are i.i.d. $N(\theta, \sigma^2)$ with σ^2 known, and consider the two simple hypotheses H_1: $\theta = \theta_1$ and H_2: $\theta = \theta_1 + \delta$.
   (a) Derive a formula for the sample size n_{NP} required to make both the Type I and Type II error probabilities equal 0.05.
   (b) Derive a formula for the sample size $n_{\mathrm{L}}(k)$ required to make both the probabilities $\Pr_1(L_1/L_2 \geq k)$ and $\Pr_2(L_2/L_1 \geq k)$ equal to 0.95.
   (c) Show that $n_{\mathrm{L}}(k) > n_{\mathrm{NP}}$ for all $k > 1$, and that this remains true if the probabilities 0.05 in (a) and 0.95 in (b) are replaced by α and $1 - \alpha$, for any $0 < \alpha < 1$.

2.2 (Continuation of Exercise 2.1)
   (a) Show that if H_1 is true then the probability of misleading strong evidence in favor of H_2 $(L_2/L_1 \geq k)$ is greatest when $n = 2(\sigma/\delta)^2 \ln(k)$.
   (b) Find the maximum probability in (a).
   (c) If H_2 is true, what is the maximum probability of (misleading) strong evidence in favor of H_1?
   (d) For the sample size in (a), what is the power of the most powerful size-α Neyman–Pearson test of H_1 versus H_2?

2.3 For independent random variables $X \sim N(\mu, 1)$ and $Y \sim N(\eta, 1)$, consider the ratio $\theta = \mu/\eta$.
   (a) Use the fact that $(X - \theta Y)/(1 + \theta^2)^{1/2}$ has a standard normal distribution to derive a 95% confidence region for θ.
   (b) Show that with positive probability the 95% confidence region will consist of the entire real line (Fieller, 1954).

2.4 Consider a model in which both the sample space and the parameter space consist of M points, $x_1, \ldots, x_M$ and $\theta_1, \ldots, \theta_M$. $P(X = x_i; \theta_i) = \alpha$ for $i = 1, \ldots, M$, with the remainder of the probability spread uniformly over the remainder of the sample space. If α is less than $1/M$, what is the best $100(1 - \alpha)\%$

confidence region $R(X)$? Give a precise interpretation of an observation $X = x$ as evidence in relation to the hypotheses H_{in}: $\theta \in R(x)$ and H_{out}: $\theta \notin R(x)$. When $\alpha = 0.05$ and $M = 19$, is the observation fairly strong evidence supporting H_{in} over H_{out}?

CHAPTER 3

Fisherian theory

3.1 Introduction

There are alternatives to the Neyman–Pearson formulation of the problem of testing statistical hypotheses. Although it is important that we recognize and understand the differences between the various formulations, there is no standard terminology to help us. Many authors have distinguished between what we are calling Neyman–Pearson tests and tests that have a different form and purpose, often calling the latter significance tests and usually citing R.A. Fisher as a particularly influential developer or proponent. Although Fisher was not the originator of significance tests, we call them 'Fisherian' because of his consistent emphasis on the distinction between the problems addressed by the Neyman–Pearson theory of hypothesis testing and problems of evidential interpretation of scientific data, for which significance tests are intended. We will draw a further distinction, describing two varieties of significance test, both of which seem to have been advocated by Fisher. The first we will call ***p*-value procedures**, and will consider in sections 3.2–3.4. These are prominent in the statistical analyses used in science. The second variety, also influential in scientific applications, we call **rejection trials**. These are particularly interesting because they link statistical hypothesis testing directly to formal logic and to the philosophy of science; they will be discussed in section 3.5.

Later in this chapter we describe how the use of significance tests to measure evidence leads to a popular evidential interpretation of confidence intervals. We also consider briefly the general issue of alternative hypotheses in science.

3.2 A method for measuring statistical evidence: the test of significance

Statistical hypothesis tests, as they are most commonly used in analyzing and reporting the results of scientific studies, do not proceed as envisioned in the Neyman–Pearson theory, with a choice between two specified hypotheses being made according to whether or not the observations fall into a pre-selected critical region. A more common procedure is described by Cox and Hinkley (1974, p. 66):

> Let $t = t(x)$ be a function of the observations and let $T = t(X)$ be the corresponding random variable. We call T a test statistic for testing H_0 if the following conditions are satisfied:
>
> (a) the distribution of t when H_0 is true is known at least approximately...
> (b) the larger the value of t the stronger the evidence of departure from H_0 of the type it is required to test...
>
> For given observations x we calculate $t_{obs} = t(x)$, say, and the *level of significance* p_{obs} by
>
> $$p_{obs} = \mathrm{pr}(T \geq t_{obs}; H_0).$$

The result of this procedure is not a decision to choose one hypothesis or another, but a number, p_{obs}, called the level of significance, or ***p*-value**; the procedure is called a **significance test**.

For example, to test the hypothesis H_0 that the probability of success is one-half on each of 20 independent trials we might use as a test statistic T the total number of successes. When H_0 is true this statistic has a known probability distribution (binomial), and large values are evidence supporting hypotheses that specify a greater success probability over H_0. If we observe 14 successes then the p-value is $\Pr(T \geq 14) = 0.06$.

An essential component of significance tests is a concept that did not appear in the Neyman–Pearson theory of hypothesis testing, the concept of strength of evidence. A p-value is supposed to indicate 'the strength of the evidence against the hypothesis' (Fisher, 1958, p. 80), with conventional interpretations as described by Burdette and Gehan (1970, p. 9):

> Reasonable interpretations of the results of significance tests are as follows:

Significance Level of Data	*Interpretation*
Less than 1 per cent	Very strong evidence against the null hypothesis
1 per cent to 5 per cent	Moderate evidence against the null hypothesis
More than 5 per cent and less than 10 per cent	Suggestive evidence against the null hypothesis
10 per cent or more	Little or no real evidence against the null hypothesis.

Another difference between hypothesis testing, in the sense of Neyman and Pearson, and significance testing is the role of alternative hypotheses: Neyman–Pearson tests are for choosing between two hypotheses, whereas significance tests are for measuring the evidence against one, the null hypothesis. Alternatives to the null hypothesis are often acknowledged to play a part in significance tests, as in Cox and Hinkley's (1974, p. 66) implicit reference to an alternative in their condition that 'the larger the value of t the stronger the evidence of *departure from H_0 of the type it is required to test*' (emphasis added). But alternative hypotheses do not have an essential explicit role analogous to the one they play in Neyman–Pearson theory. In fact, many authorities maintain that significance tests' freedom from dependence on explicit alternative hypotheses is essential in some important applications (such as 'goodness of fit' tests):

> Let us try the simple single hypothesis first. If the data do not fit that, then it is worth while going ahead [and constructing alternative hypotheses]. If it is consistent with the data let us not waste our time. (Barnard, in Savage, 1962, p. 85)

Box (1980) has defended this position more recently.

Here is a summary of some of the differences between these two approaches to testing hypotheses about the distribution of a random variable X:

Neyman–Pearson hypothesis tests	*Significance tests (p-value procedures)*
Purpose:	
To choose one of two specified hypotheses, H_1 and H_2, on the basis of an observation $X = x$.	For a single hypothesis H, to measure the evidence against H represented by an observation $X = x$.
Elements:	
1. Two hypotheses (families of probability distributions) H_1 and H_2.	1. One hypothesis H, called the 'null' hypothesis.
2. A test function $\delta(x)$ that specifies which hypothesis to choose when $X = x$ is observed: if $\delta(x) = 1$ we choose H_1, if $\delta(x) = 2$ we choose H_2.	2. A real-valued function $t(x)$ that gives an ordering of sample points as evidence against H: $t(x_1) > t(x_2)$ means that x_1 is stronger than x_2 as evidence against H.

3. Result is a decision or action, 'Choose H_1' or 'Choose H_2'.

3. Result is a number, the significance level, or p-value, interpreted as a measure of the evidence against H; the smaller the p-value the stronger the evidence.

The distinction between Neyman–Pearson tests and significance tests is not made consistently clear in modern statistical writing and teaching. Mathematical statistical textbooks tend to present Neyman–Pearson theory, while statistical methods textbooks tend to lean more towards significance tests. The terminology is not standard, and the same terms and symbols are often used in both contexts, blurring the differences between them. For example, descriptions of Neyman–Pearson theory often refer to the size, or Type I error probability, as the 'significance level'.

A further source of confusion is that within the Neyman–Pearson framework it is sometimes recommended that the experimenter should report, not the result of testing H_1 versus H_2 at a preselected Type I error level α, but the smallest value of α that would have led to rejection of H_1. This enables the reader who prefers a different Type I error level, say α', to perform his own test, rejecting H_1 (choosing H_2) if the reported α is smaller than his α'. Such a reported α is mathematically equivalent to a p-value (and is sometimes called by that name). But this does not make the procedure into a significance test, which is defined, not simply by what number is calculated, but by what that number is supposed to mean. As we saw in sections 2.3 and 2.4, Neyman was quite right in his insistence on a narrow behavioral, or decision-making, interpretation of his theory – evidential interpretations are generally invalid.

The key difference between Neyman–Pearson tests and significance tests is in their purpose. Neyman–Pearson tests are rules for choosing between alternative actions, while significance tests purport to measure evidence. That is, Neyman–Pearson tests address the second of the physician's three questions in Chapter 1, 'What should I *do*?', while significance tests address the third, 'How should I interpret these observations as *evidence*?'. In his section on 'The simple test of significance', Fisher (1956, p. 42) complained that the Neyman–Pearson view 'that the purpose of the test is to discriminate or "decide" between two or more hypotheses' had 'greatly obscured' the understanding of tests.

He then offered

> a clear view of the nature of a test of significance applied to a single hypothesis by a unique body of observations.
>
> Though recognizable as a psychological condition of reluctance, or resistance to the acceptance of a proposition, the feeling induced by a test of significance has an objective basis in that the probability statement on which it is based is in fact communicable to, and verifiable by, other rational minds. The level of significance in such cases fulfils the conditions of a measure of the rational grounds for the disbelief it engenders.

3.3 The rationale for significance tests

Why should a small p-value be interpreted as signifying strong evidence against the hypothesis? Barnard (1967, p. 32) explains:

> The meaning of 'H is rejected at significance level α' is 'Either an event of probability α has occurred, or H is false,' and our disposition to disbelieve H arises from our disposition to disbelieve in events of small probability.

This echoes Fisher's (1959, p. 39) explanation – after calculating, under a random distribution hypothesis, that the probability of the event observed, or a more extreme event, was about 1/33 000, he proposed that this probability

> is amply low enough to exclude at a high level of significance any theory involving a random distribution.
>
> The force with which such a conclusion is supported is logically that of the simple disjunction: *Either* an exceptionally rare chance has occurred, *or* the theory of random distribution is not true.

According to the Fisher–Barnard explanation, significance tests rest on some principle like the following:

> *Law of improbability*: If hypothesis A implies that the probability that a random variable X takes on the value x is quite small, say $p_A(x)$, then the observation $X = x$ is evidence against A, and the smaller $p_A(x)$, the stronger that evidence.

More recently Cox (1977, p. 53) has cited this law ('the smaller is the probability under H_0, the stronger is the evidence against H_0') as the basis for significance tests in some circumstances. But the law of improbability has attracted criticism as well as support. Some have observed that it appears to be unacceptably hard on null hypotheses. Suppose, for example, that we have a computer program intended to generate standard normal deviates. Consider the null hypothesis that the program is operating correctly. Now consider the evidence in a single observed output, $X = x$. Because the hypothesis implies that the probability of any single point is zero, the law of improbability would imply that whatever value, x, is produced, it is overwhelming evidence that the program is not working properly. The problem is not restricted to continuous distributions – if X is intended to have a $Bin(n, \frac{1}{2})$ distribution then the maximum probability on any outcome is roughly $(2/n\pi)^{1/2}$, so that if n is large then no matter what value x is observed, it will be judged to be strong evidence against the (true) binomial distribution hypothesis. This is the point that Hacking (1965, p. 82) made in discussing Fisher's argument quoted above: 'if Fisher's disjunction had any force, we should always have to exclude any hypothesis like that of random distribution, whatever happened. So it has no force'.

The binomial distribution assigns greater probability to values of x near $n/2$. Although the absolute probabilities are all small when n is large, the relative probabilities are not, and the ratio of the maximum probability to the minimum, which occurs at both $x = 0$ and at $x = n$, is quite large, roughly $2^n(2/n\pi)^{1/2}$. Thus although all possible outcomes have low probability under the hypothesis, some have much lower probabilities than others. To accommodate this phenomenon, we might try a modified version of the law stating that it is low probability *relative to other outcomes* that makes a given outcome evidence against a hypothesis.

> *Law of improbability II*: If hypothesis A implies that the probability that a random variable X takes on the value x is small compared to the probability of another value x', $p_A(x) \ll p_A(x')$, then the observation $X = x$ is evidence against A, and the smaller the ratio $p_A(x)/p_A(x')$, the stronger the evidence.

Law II is unsatisfactory on various counts, one of which is that it leaves some important hypotheses exempt from unfavorable evidence. Suppose X represents a series of n Bernoulli (success or failure, coded 1 or 0) trials, and consider the hypothesis that the trials are independent with common probability of success equal to one-half. Under this hypothesis every possible outcome is a series of n zeroes and ones, and they all have the same probability of occurrence, $(\frac{1}{2})^n$. Thus for every pair of possible outcomes, x and x', $p_A(x)/p_A(x') = 1$, indicating evidence of no strength at all; no outcome is less probable than any other, so none is evidence against the hypothesis.

Maybe we need to bring the 'more extreme' outcomes into the analysis. Since the occurrence of an event whose probability under H is small is interpreted as evidence against H, with the strength of evidence growing as the probability shrinks, the outcomes that are 'as extreme or more so' are apparently just those outcomes whose probabilities under H are as small or smaller. Suppose we try to state the law in terms of the probabilities of outcomes 'as extreme or more so' than the one observed:

> *Law of improbability III*: If hypothesis A implies that the probability that a random variable X takes on the value x is $p_A(x)$ and if the sum $S(x)$ of the probabilities of all values whose probabilities are less than or equal to $p_A(x)$ is small, then the observation $X = x$ is evidence against A, and the smaller the sum $S(x)$, the stronger the evidence.

But law III also fails in the simple case of a sequence of independent Bernoulli trials with success probability one-half. Since all possible outcomes have the same probability, $p_A(x) = (\frac{1}{2})^n$, for every one $S(x) = 1$, again indicating evidence of no strength at all. According to law III only outcomes that are *impossible* under this null hypothesis are evidence against it.

We will not continue to fiddle with the law of improbability, trying to adjust our statement of it until we get it right. It cannot be made right, as we already learned in section 1.4: it is not low probability under A that makes an event evidence against A – it is low probability under A relative to the probability under another hypothesis B that makes it evidence supporting B over A. And then it is not evidence against A, but evidence against A, *vis-à-vis* B.

Suppose I send my valet to bring my urn containing 100 balls, of which only two are white. I draw one ball and find that it is white. Is this evidence against the hypothesis that he has brought the correct urn? And is $p = 0.02$ a proper measure of the strength of this evidence? Suppose that I keep in my urn vault two urns, one with two white balls and another, identical in appearance, that contains no white balls. Now is my observation of a white ball evidence that he has not brought the right urn? Fisher's disjunction still applies – either a rare event has occurred or the null hypothesis (correct urn) is false. But although the observation of a white ball is rare under the null hypothesis, it is even rarer under the alternative (wrong urn). In this case, the observation is actually strong evidence *in favor* of the null hypothesis. Of course, we might consider other hypotheses as well. For example, if my valet likes to play tricks, we might consider the hypothesis that he has added some more white balls to the urn. The evidence favors that hypothesis over the 'correct urn' null hypothesis by a factor that depends on how many white balls he might have added.

The point again is that evidence is relative (as we saw in section 1.4) – whether it counts for or against one hypothesis can only be determined with reference to an alternative (see Exercise 3.1). This point has been made well and often for decades. Before the birth of the Neyman–Pearson theory the inventor of the *t*-test, W.S. Gosset, explained to Neyman's coauthor, Egon Pearson, that an observed discrepancy between a sample mean and a hypothesized population mean

> doesn't in itself necessarily prove that the sample was not drawn randomly from the population even if the chance is very small, say .00001: what it does is to show that if there is any alternative hypothesis which will explain the occurrence of the sample with a more reasonable probability, say .05 . . . you will be very much more inclined to consider that the original hypothesis is not true.
>
> (Gosset [1926], quoted in Pearson, 1938)

So the Fisher–Barnard rationale for significance tests, as expressed in the law of improbability, is wrong. There is, in fact, no sound rationale for these tests. This is because they are incompatible with the law of likelihood. Specifically, significance tests depend critically on how the probability distribution is spread over unobserved points in the sample space (through their definition in terms of outcomes 'as extreme or more so' than the one observed) and are therefore incompatible with the law of likelihood's implication of the 'irrelevance of the sample space' (section 1.11). This point is pursued in the next section, where a conspicuous problem with the interpretation of significance tests is also described. The existence of such problems supports the above claim that the reason why a plausible rationale for significance tests has not yet been found is that none exists.

3.4 Troubles with *p*-values

Let us look at the role of outcomes 'as extreme or more so' in significance tests. In problems where there is a well-defined alternative hypothesis, we can certainly identify such outcomes: if f_1 and f_2 are densities corresponding to the null and alternative distributions respectively, and if x_0 is the observation, then the set

$\{x; f_2(x)/f_1(x) \geq f_2(x_0)/f_1(x_0)\} \equiv S(x_0)$ consists of all the outcomes that are 'as extreme or more so' compared to x_0. These are the outcomes that would give a likelihood ratio supporting H_2 over H_1 as great as or greater than the ratio associated with x_0.

The *p*-value, $\Pr_1(S(x_0))$, consists not only of the probability of what was observed (x_0), but also of the probabilities of all the more extreme outcomes that did not occur. But a proper measure of strength of evidence should not depend on the probabilities of unobserved values. To see this, recall the example in section 1.10, where 20 tosses were made with a coin whose probability of heads (success), θ, is unknown. The result is reported in a code that is known to you; I, on the other hand, know only the code word for '6'. If the number of heads observed is six, then you and I obtain precisely the same evidence about θ. Thus if we both consider $H_1: \theta = 0.5$ and an alternative asserting that the proportion is somewhat lower, say $H_2: \theta = 0.3$, then your prior probability ratio, $\Pr(H_2)/\Pr(H_1)$, and mine will both be increased by the same factor, 5.18. But our *p*-values for testing H_1 versus H_2 do not agree. Yours is $p_1(X = 6) + p_1(X = 5) + \ldots + p_1(X = 0) = 0.06$. On the other hand, since I can observe only '6' or 'not-6', the observed outcome is the most extreme possible one for me, and my *p*-value is just its probability, $p_1(X = 6) = 0.04$. The *p*-values assert (incorrectly) that the outcome (six heads in 20 tosses) is stronger evidence against H_1 (in favor of H_2) for me than it is for you.

The significance-test approach to measuring the evidence is wrong because its dependence on the sample space leads to different answers in situations where the evidence is the same. That is, it violates the principle of the 'irrelevance of the sample space' (section 1.11). This becomes even clearer if we provide some more details about this experiment. It turns out that you have memorized only the code-word for '6'. If any other result had occurred, you would have had to consult your code-book to find how many heads had been observed. Now, long after the experiment has been completed and the *p*-values have been published, we are storing some bent coins in your vault and we happen to notice that your code-book is missing.

So your situation was actually the same as mine – if $X = 4$ had occurred you could have recognized it only as 'not-6'. Therefore your sample space was the same as mine, {6, not-6}, and your calculated *p*-value, 0.06, is wrong. You conscientiously draft a letter to the journal where your result was published, apologizing for your error and reporting the corrected *p*-value, 0.04. But then your secretary, when he sees the letter, sheepishly confesses that he threw away the code-book while tidying up the vault. Now the plot thickens. If he threw the book away *before* the outcome 'six successes in 20 tosses' was reported, then the appropriate p-value is the corrected one; but if the clean-up took place *later*, so that the code-book was still available when it might have been needed (but was not), then your sample space was $\{0, 1, \ldots, 20\}$ after all, and so your original p-value is still valid.

This is clearly silly – for the data actually observed, for the evidence actually obtained, the code-book was not needed. The evidence about θ, unlike the p-value, does not depend on when the book disappeared – that is, it does not depend on which sample space, $\{0, 1, \ldots, 20\}$ or $\{6, \text{not-}6\}$, you were sampling from when $X = 6$ was observed. (This example is a descendant of one constructed by Pratt (1961) that has figured prominently in modern discussions of the foundations of statistical inference.)

There is a significant piece of indirect evidence that something is seriously wrong with significance tests. According to the widely used 'Reasonable interpretations of the results of significance tests' described by Burdette and Gehan, and quoted earlier, a given p-value has a more or less fixed meaning. For example, a p-value between 1% and 5% is supposed to indicate 'moderate evidence against the null hypothesis'; a value less than 1% indicates 'very strong evidence'. This concept, that equal p-values represent equal amounts of evidence, at least approximately, was named the 'α-postulate' by Cornfield (1966). Fisher (1934, p. 182) states it as follows:

> It is not true... that valid conclusions cannot be drawn from small samples; if accurate methods are used in calculating the probability [the p-value], we thereby make full allowance for the size of the sample, and should be influenced in our judgement only by the value of the probability indicated.

Berkson's (1942) statement was only slightly less forceful: 'the evidence provided by a small p correctly evaluated is broadly independent of the number in the sample'. The central role of significance tests in many research areas rests on the α-postulate – results with a p-value between 0.01 and 0.05 are flagged with an asterisk and declared to be 'statistically significant', while those with a p-value smaller than 0.01 are given two asterisks and declared 'highly significant'. The acceptability of a research report for publication often depends on whether key results are 'significant' or not.

But the α-postulate is wrong. In their preface to the *New Cambridge Elementary Statistical Tables*, Lindley and Scott (1984, p. 3) explain:

> the interpretation to be placed on the phrase 'significant at 5%' depends on the sample size: it is more indicative of the falsity of the null hypothesis with a small sample than with a large one.

Thus a given p-value does not have a fixed meaning. If two experiments that are identical except for their sample sizes produce results with the same p-value, these results do not represent equally strong evidence against the null hypothesis – the evidence is stronger in the smaller experiment.

Ten of the world's most influential applied statisticians co-authored a paper in which they, too, explained that the α-postulate is false: 'A given p-value in a large trial is usually stronger evidence that the treatments really differ than the same p-value in a small trial of the same treatments would be' (Peto *et al.*, 1976, p. 593). Their interpretation is opposite that of Lindley and Scott.

Does a significance level of $p = 0.04$ indicate 'moderately strong' evidence against the null hypothesis, regardless of sample size, as the α-postulate and common practice imply? Or does it indicate stronger evidence in a small sample than in a large one, as Lindley and Scott state? Or does it indicate stronger evidence in a large sample as Peto *et al.* assert?

We should not be surprised to find that a statistical procedure that purports to measure evidence, but in a way incompatible with the law of likelihood, is mired in paradox and controversy (Royall, 1986; see also Morrison and Henkel, 1970).

3.5 Rejection trials

We have contrasted two ways to formulate statistical hypothesis-testing problems. The one developed by Neyman and Pearson addresses problems of choosing between two hypotheses, avoiding our central question of how to interpret statistical data as evidence. The other aims directly at out target – it seeks to measure the strength of evidence – but misses the mark. Neither of these formulations seems to capture the spirit of the definition given in the textbook from which many of today's practicing statisticians learned the basics:

> A procedure which details how a sample is to be inspected so that we may conclude that it either agrees reasonably with the hypothesis or

> does not agree with the hypothesis will be called a test of the hypothesis. (Dixon and Massey, 1969, p. 76)

Here a hypothesis test is seen as a decision procedure, *à la* Neyman–Pearson theory, but with some important differences. Although there are two possible conclusions or 'actions', only one hypothesis is mentioned. And the phrase 'either agrees reasonably with the hypothesis or does not' suggests that the two conclusions correspond to definite evidential interpretations of the sample. In many scientific applications of statistical tests a similar view is adopted. While the objective is a rule or procedure for choosing between two alternatives, as in the Neyman–Pearson paradigm, the two alternatives are now stated in terms of a single hypothesis – one is favorable to the hypothesis and the other unfavorable. And an essential part of the reasoning is that choosing the unfavorable conclusion is justified only when the sample represents sufficiently strong evidence against the hypothesis (as in the Dixon and Massey scenario when the sample does not agree 'reasonably' with the hypothesis).

This third formulation views statistical hypothesis testing as a process analogous to testing a proposition in formal logic via the argument known as *modus tollens*, or 'denying the consequent': if A implies B, then not-B implies not-A. We can test A by determining whether B is true. If B is false, then we conclude that A is false. But, on the other hand, if B is found to be true we cannot conclude that A is true. That is, A can be proven false by such a test, but it cannot be proven true – either we disprove A or we fail to disprove it. (This is the form of argument that is used in mathematics when a false proposition is disproved by a counterexample.) When B is found to be true, so that A survives the test, this result, although not proving A, does seem intuitively to be evidence supporting A. Whether this evidential interpretation is correct or not is the subject of Hempel's famous 'paradox of the ravens', which is discussed in the Appendix. This form of reasoning is at the heart of the philosophy of science, according to Popper (see Putnam, 1974). Its statistical manifestation is in this third formulation of hypothesis testing that we will call 'rejection trials'.

In applications of this third form of testing, a statistical hypothesis H_0, the 'null' hypothesis, plays a role analogous to that of the proposition A in that it can be disproved but not proved, rejected but not accepted (Noether, 1971, p. 64). Fisher (1966, section II.8) explains:

> it should be noted that the null hypothesis is never proved or established, but is possibly disproved, in the course of experimentation. Every experiment may be said to exist only in order to give the facts a chance of disproving the null hypothesis.... The notion of an error of the so-called 'second kind,' due to accepting the null hypothesis 'when it is false'... has no meaning with reference to simple tests of significance...

The experimenter identifies a rejection region R that has small probability under H_0, so that, if H_0 is true, then the event 'X is not in R' has high probability. This event, call it E, has a role analogous to that of the proposition B in the *modus tollens* argument. The analogy is imperfect because whereas A implied B (with perfect certainty), H_0 implies E with high probability; that is, if A is true then not-B is impossible, while if H_0 is true then not-E (X in R) is merely improbable. But the form of reasoning in the statistical version of the problem parallels that in deductive logic: if H_0 implies E (with high probability) then not-E justifies rejecting H_0.

The term 'test of significance' is often used, as it was in section 3.2, to refer to procedures that produce p-values for measuring the evidence against a hypothesis. The same term is also used, as it was in the above quote from Fisher's *Design of Experiments*, to refer to the (reject/do not reject) procedures just described. This latter usage seems more apt – here the hypothesis is subjected to a test. If it fails the test, it is rejected; if not, it survives, perhaps to face another test. We use the terms 'p-value procedures' and 'rejection trials' to distinguish between these two visions of 'tests of significance'.

Rejection trials are similar to Neyman–Pearson procedures in some respects. First, their objective is stated not as measuring the evidence against the hypothesis, but as choosing between two alternative actions, that is, choosing whether to reject the hypothesis or not. And like Neyman–Pearson tests, they require that an error rate α be selected and that a rejection region having probability no greater than α under H_0 be determined. Then H_0 is rejected if the observed value of the random variable falls in the region.

An important difference between rejection trials and Neyman–Pearson tests becomes clear when the observation does not fall in the rejection region: Neyman–Pearson tests require choosing between two alternatives, H_0 and H_1, so that the complement of the region where H_0 is rejected (and H_1 accepted) is the region where H_1 is rejected (and H_0 accepted). Rejection trials, on the other hand, are viewed as challenges to the single null hypothesis H_0. If the observation is in the rejection region, then H_0 fails the challenge and is rejected; otherwise the result is 'Do not reject H_0'. That is, the symmetry described by Neyman (1950, p. 259), 'it is immaterial which of the two alternatives...is labelled the hypothesis tested', is clearly missing in the trials described by Fisher (1966, section II.8) 'in which the only available expectations are those which flow from the null hypothesis being true'. Thus Fisz (1963, p. 426) writes: 'In general, a significance test [rejection trial] allows us to make decisions only in one direction'. If the observation is in the rejection region 'then H_0 may be rejected', but if not 'then we can only state that the experiment does not contradict H_0'.

We are concerned in this monograph with how statistical data are interpreted as evidence. From this viewpoint the key difference between Neyman–Pearson tests and rejection trials is not in the existence, explicit or not, of an alternative statistical hypothesis, nor in the relationship between such an alternative and the null hypothesis. The key difference is that, unlike Neyman–Pearson tests, rejection trials entail evidential interpretation of the observations. In these trials the rejection of H_0 is justified when x falls in the rejection region, it is said, because such observations 'do not agree with' or 'do not fit' the hypothesis; they 'are inconsistent with', 'contradict', or even 'disprove' it. If under H_0 the probability of the rejection region is α, then the observations are said 'to provide sufficient evidence to cause rejection', or to be 'statistically significant' at level α. Whatever expression is used, the implication is that observations in the rejection region are evidence against the hypothesis; and observations in a rejection region with very small α are very strong evidence.

In section 2.3 we considered an example of Cox (1958) in which a coin toss is used to determine whether one or k i.i.d. $N(\theta, \sigma^2)$ observations will be made. There we looked at confidence intervals for θ. However, Cox's original example was stated in terms of hypothesis tests, and it dramatizes the difference between Neyman–Pearson tests and significance tests of the 'rejection-trial' variety. For simplicity let $\sigma^2 = 1$, and suppose the sample size when the coin falls tails is $k = 100$. The hypotheses are H_0: $\theta = 0$ and H_1: $\theta = 1$. Cox (1958) observed that if instead of using the coin toss we choose the sample size – say, n – deliberately, then the best (most powerful) test of size $\alpha = 0.05$ is to reject H_0 if and only if $\bar{x} > 1.645/\sqrt{n}$. So again we consider procedure A: if the coin falls heads, so $n = 1$, and $X = x$ is observed, reject H_0 if $x > 1.645$; if it falls tails and 100 observations are made, reject H_0 if $\bar{x} > 1.645/10$. Procedure A consists of using, for each sample size, 1 and 100, the best test of size 0.05. It has power $\frac{1}{2} \times 0.259 + \frac{1}{2} \times 1.000 = 0.63$. But again we can do better. The most powerful test of size 0.05 is given by procedure B: reject H_0 if the coin falls heads and $x > 1.282$ or if it falls tails and $\bar{x} > 5.078/10$. Procedure B's Type I error rate is $\frac{1}{2} \times 0.100 + \frac{1}{2} \times 0.000 = 0.05$, the same as A's, but its power is greater: $\frac{1}{2} \times 0.389 + \frac{1}{2} \times 1.000 = 0.69$.

For one whose problem is accurately represented by the Neyman–Pearson formulation, one who truly seeks to minimize the Type II error rate subject to the constraint that the Type I rate not exceed 0.05, it might come as a surprise that A is not the better procedure. But B's superiority, though surprising, is real, and there is no reason to prefer A. On the other hand, if the rejection-trial formulation is more apt, procedure B is not better – in fact, it is widely considered to be quite wrong. Cox (1958), calling procedure A the conditional test and B the unconditional one, wrote

> Now if the object of the analysis is to make statements by a rule with certain specified long-run properties, the unconditional test...is in order. ...If, however, our objective is to say what we can learn from the data we have, the unconditional test is surely no good. Suppose that we know we have [only one] observation...The unconditional test says that we can assign this a higher level of significance than we ordinarily do, because if we were to repeat the experiment, we might sample some quite different distribution [i.e. we might make 100 observations instead of only one].

Procedure B is 'no good' because when only one observation is made it rejects at the 5% level whenever $X > 1.282$, and this is evidently too liberal – to properly claim 5% significance we should require, as A does, $X > 1.645$. Procedure B compensates 'on the average' by being overly conservative when $n = 100$, rejecting H_0 at the 5% significance level only on the basis of quite extreme outcomes, $\sqrt{n}\bar{X} > 5.078$. From the significance-testing viewpoint, Procedure B will not do because the objective is to characterize the evidence properly in each case; B allows the claim of 5% significance on the basis of evidence that is too weak when $n = 1$, requiring evidence that is too strong when $n = 100$.

3.6 A sample of interpretations

The distinctions between the three views of hypothesis testing that we have considered are useful for understanding the rationale and interpretation of statistical tests. It is quite possible, however, that none of the three is a precise representation of what any one statistical author means by 'hypothesis testing'. The following quotations certainly do not represent a single viewpoint. Instead each author describes a slightly different vision, each drawing elements from all of the three formulations that we have tried to distinguish. But the point of view that we have called 'rejection trials' is influential in each description.

> In the testing process the null hypothesis either is rejected or is not rejected. If the null hypothesis is not rejected, we will say that the data on which the test is based do not provide sufficient evidence to cause rejection. (Daniel, 1991, p. 192)

> A nonsignificant result does not prove that the null hypothesis is correct – merely that it is tenable – our data do not give adequate grounds for rejecting it. (Snedecor and Cochran, 1980, p. 66)

> The verdict does not depend on how much more readily some other hypothesis would explain the data. We do not even start to take that question seriously until we have rejected the null hypothesis.
>
> ... The statistical significance level is a statement about *evidence*... If it is small enough, say $p = 0.001$, we infer that the result is not readily explained as a chance outcome if the null hypothesis is true and we start to look for an alternative explanation with considerable assurance.
>
> (Murphy, 1985, p. 120)

> If [the p-value] is small, we have two explanations – a rare event has happened, or the assumed distribution is wrong. This is the essence of the *significance test* argument. Not to reject the null hypothesis... means only that it is accepted for the moment on a provisional basis.
>
> (Watson, 1983)

> *Test of hypothesis.* A procedure whereby the truth or falseness of the tested hypothesis is investigated by examining a value of the test statistic computed from a sample and then deciding to reject or accept the tested hypothesis according to whether the value falls into the critical region or acceptance region, respectively.
>
> (Remington and Schork, 1970, p. 200)

> Although a 'significant' departure provides some degree of evidence against a null hypothesis, it is important to realize that a 'nonsignificant' departure does not provide positive evidence *in favour* of that hypothesis. The situation is rather that we have failed to find strong evidence against the null hypothesis. (Armitage and Berry, 1987, p. 96)

> If that value [of the test statistic] is in the region of rejection, the decision is to reject H_0; if that value is outside the region of rejection, the decision is that H_0 cannot be rejected at the chosen level of significance ... The reasoning behind this decision process is very simple. If the probability associated with the occurrence under the null hypothesis of a particular value in the sampling distribution is very small, we may explain the actual occurrence of that value in two ways; first we may explain it by deciding that the null hypothesis is false or, second, we may explain it by deciding that a rare and unlikely event has occurred. (Siegel and Castellan, 1988, Chapter 2)

3.7 The illogic of rejection trials

The above quotes suggest that the rejection trial is a method for determining when a given set of observations represents sufficiently strong evidence against a hypothesis to justify rejecting that hypothesis. But when it is given this interpretation the method defies the rules of logic.

Consider the $Bin(n, \theta)$ model for X and the hypothesis H_0: $\theta = \frac{1}{2}$. When the observed value is x, we are justified in rejecting H_0 at level α if $\Pr_0(X \geq x) \leq \alpha/2$. If, on the other hand, we are testing the hypothesis H_0': $\theta \leq \frac{1}{2}$, our observation x is strong enough evidence to justify rejecting if $\Pr_0(X \geq x) \leq \alpha$. Thus a value x for which $\alpha/2 < \Pr_0(X \geq x) \leq \alpha$ represents strong enough evidence to justify rejecting the composite hypothesis that either $\theta = \frac{1}{2}$ or $\theta < \frac{1}{2}$, but it is not strong enough evidence to justify rejecting the simple hypothesis that $\theta = \frac{1}{2}$. We may conclude (at significance level α) that both $\theta = \frac{1}{2}$ and $\theta < \frac{1}{2}$ are false, but we may not conclude that $\theta = \frac{1}{2}$ alone is false. We may conclude 'neither *A* nor *B*' but we may not conclude 'not-*A*'. Odd.

This interpretation of rejection trials makes no more sense if it is expressed in terms of the alternatives to the hypotheses tested. If, when we reject $\theta = \frac{1}{2}$, we are concluding that either $\theta < \frac{1}{2}$ or $\theta > \frac{1}{2}$, then clearly this is justified by any evidence that justifies the stronger conclusion that $\theta > \frac{1}{2}$. That is, if the evidence justifies the conclusion that *A* is true, then surely it justifies the weaker conclusion that either *A* or *B* is true. Rejection trials do not conform to this logic.

3.8 Confidence sets from rejection trials

Rejection trials provide the basis for an evidential approach to defining and interpreting confidence sets. If we have for each possible value of a parameter θ a level-α test of significance (rejection trial) of the hypothesis that the parameter equals that value, then we can define a $100(1-\alpha)\%$ confidence set. This set consists simply of all the values of θ that would not be rejected by the corresponding test. That is, if the hypothesis H_0: $\theta = \theta_0$ is not rejected on the basis of the observation $X = x$, then θ_0 is in the set $S(x)$. That this procedure does indeed produce a $100(1-\alpha)\%$ confidence set follows directly from the fact that for every θ_0 the random set $S(X)$ includes θ_0 if and only if a value of X is observed which does not lead to the rejection of H_0: $\theta = \theta_0$, and the probability of this, when H_0 is true, is at least $1-\alpha$. Thus $\Pr_\theta(S(X)$ will include $\theta) \geq 1-\alpha$ for every θ, which is to say, $S(X)$ is a valid $100(1-\alpha)\%$ confidence-set procedure.

This approach gives an explicit evidential interpretation to the confidence set, which now consists of all the values of θ that are consistent with the observation $X = x$ in the sense that this observation would not justify their rejection at significance level α. Values excluded from the confidence set are those against which $X = x$ represents evidence strong enough to warrant rejection at level α.

This interpretation is sometimes invoked in order to 'make sense' of a confidence set that seems paradoxical when interpreted in terms of one's confidence that it contains the true parameter value. A popular example is the confidence set for a ratio of two normal means (Exercise 2.3). The 95% confidence set can turn out to be the whole real line. Since this set contains all possible values of the ratio, it seems ridiculous to assign to it a confidence coefficient of only 0.95 – we are actually 100% confident that it contains the true ratio of means. The rejection-trial interpretation is attractive: the confidence set excludes only those values against which we have sufficiently strong evidence to justify rejection of the corresponding hypothesis at the 5% level. Now in this example the samples that give the entire line as the confidence set are those in which the estimates of both numerator and denominator are very close to zero. Such samples tell us very little about the ratio; as Exercise 7.6 shows, they represent only weak evidence. They do not justify our rejecting any of the possible values of the ratio. All of the values are 'consistent with the observations at the 5%

level', and this is what the (very large) confidence region correctly shows.

The evidential interpretation of confidence sets that is provided by the significance-testing (rejection-trial) approach is attractive. But it is valid only if the evidential interpretation of rejection trials is valid. And this is not the case, because the rationale for rejection trials is the same as that for p-value procedures – it rests on Fisher's disjunction, as explained by Watson and by Siegel and Castellan in the quotations in section 3.6. Rejection trials fail, as tools for evidential interpretation of statistical data, for the same reasons that p-value procedures fail. Rejection trials lead to different answers in situations where the evidence is the same, just as p-value procedures were shown to do in section 3.4. In terms of the urn example discussed there, whether the coded report of six successes in 20 tosses of the bent coin is or is not 'significant at the 5% level' for testing $H_0\colon \theta = \frac{1}{2}$ depends on whether the code-book would have been available if a different number of successes had occurred. The immediate problem is the dependence of the significance test procedures, of both the p-value and the rejection-trial varieties, on the sample space. The underlying reason, explained in section 3.3, is that the law of improbability is not tenable.

3.9 Alternative hypotheses in science

As we discussed in section 3.3, the law of likelihood applies to pairs of hypotheses and suggests that a sound theory of evidence in relation to a single statistical hypothesis is impossible. Unfortunately, the use of significance-testing methodology has trained many scientists as well as statisticians to think in terms of evidence against single hypotheses, as illustrated in the quotations in section 3.6. Since the problem can be formulated in terms of one hypothesis and a test statistic (as in the description by Cox and Hinkley in section 3.2), with no explicit alternative required, it is easy to overlook the essential role played by alternative hypotheses.

Are there statistical 'null' hypotheses that are scientifically important? If so, they are rare. The reason is the familiar observation that our statistical models are only approximations to real-world phenomena and processes. The answer to the question 'Is the null hypothesis correct?' is always the same – no! Does the odds ratio equal 1? No. Does the regression coefficient equal zero? No. Are the two distributions identical? No. If the purpose of experiments were to answer such questions, there would be no point in doing experiments, since we already know the answers.

Experiments like the following are sometimes cited as counter-examples to the above claim. To test whether a subject is capable of extrasensory perception (ESP), a random sequence of images is generated but concealed from the subject. The images may be the cards in a well-shuffled deck or a sequence of zeroes and ones generated by a process such as tossing a coin. The subject is asked to reproduce the sequence. Early experiments of this sort were plagued by the possibility that subjects were given inadvertent cues to the correct responses (via normal sensory channels) or were able to cheat (Hansel, 1966). Let us assume that we can eliminate these flaws in the experimental setup. If the subject has no ESP ability then the number of terms that he correctly matches has a simple probability distribution that becomes the null hypothesis. Any departure from that distribution would show ESP ability. For simplicity, suppose the images are generated by a sequence of independent Bernoulli trials with probability $\theta = \frac{1}{2}$. If the subject's success probability is anything different from $\frac{1}{2}$, this is taken to reflect ESP. If his probability is truly greater than $\frac{1}{2}$, this clearly means that he is receiving some extrasensory information. But a probability less than $\frac{1}{2}$ means the same (and that he is misinterpreting the information). Any departure from the null hypothesis that his number of successes in n trials has a $Bin(n, \frac{1}{2})$ probability distribution proves the existence of ESP. It seems that we really do want to answer the question 'Is the null hypothesis true?'. If it is not, then ESP exists.

The problem, of course, is that no one can generate a perfect sequence of i.i.d. $Bernoulli(\frac{1}{2})$ trials. Certainly it cannot be done by tossing a coin, for all coins are imperfect and the probability of heads is never exactly one-half. Likewise, the subject who has no ESP ability, but is simply guessing, cannot produce a perfect sequence of i.i.d. $Bernoulli(\frac{1}{2})$ guesses. Then there is always some probability of error in recording and transmitting the results. This means that the null hypothesis is always false, whether or not the subject has ESP ability. The $Bin(n, \frac{1}{2})$ probability distribution is only an imperfect model for the number of matches observed in n trials.

The key question then becomes 'Does the probability distribution differ from the $Bin(n, \frac{1}{2})$ by more than can be reasonably explained in terms of the inevitable imperfections in the mechanism for generating the sequence of images, checking for matches, and recording the results?'. This question refers not only to the null hypothesis but also to alternatives. Results leading to rejection of the null hypothesis at a very small p-value do not necessarily represent evidence for ESP. If $n = 100$ million and $x = 50.02$ million successes are observed then $2\sqrt{n}(\bar{x} - 0.5) = 4.0$, giving a very small p-value, 0.000 03. These observations are quite strong evidence for a success probability of 0.5002 versus 0.5000 ($LR > 2900$). But a difference this small, an excess of two expected successes per 10 000 trials, might well be explained in terms of imperfections in the experiment, and at any rate would appear to represent the absence of an empirically meaningful ESP phenomenon.

The meaningful question, as explained by Gossett in the quote in section 3.3, is not 'Are the observations evidence against the null hypothesis?' but 'Are there scientifically meaningful alternative hypotheses that are better supported?'.

3.10 Summary

Today's statistical practice is directed by an informal blending of Neyman–Pearson theory with concepts and interpretations that are not a part of that theory. We call this approach Fisherian. Scientific applications of hypothesis testing, for example, are usually of a type so different from the procedures described by Neyman–Pearson theory that they are given a special name, tests of significance. There are actually two distinct types of significance test, namely p-value procedures and rejection trials. Both explicitly attempt to do what Neyman–Pearson theory does not – to quantify the strength of statistical evidence. Significance tests fail in this endeavor because they rest on the faulty foundation of the law of improbability. Fisherian methods in general, as tools for representing and interpreting statistical data as evidence, fail for the same reason – they rest on the law of improbability and violate the law of likelihood.

Exercises

3.1 (a) Suppose you observe a random variable X and are interested in the simple hypothesis H_0: $X \sim Bin(100, 0.5)$. Is the observation $X = 37$ strong evidence against H_0? How about $X = 50$? Explain. [Some numbers that you might want to consider are: $Pr_0(X = 37) = 0.003$, $Pr_0(X \leq 37) = 0.006$; $Pr_0(X = 50) = 0.080$, $Pr_0(X \leq 50) = 0.540$.]

(b) Now suppose you learn that X was produced by making 100 draws from an urn containing 100 balls, 50 black and 50 white, and counting the number of draws on which a black ball was seen. The hypothesis H_0 in (a) is true if the draws were made *with* replacement. Is the observation $X = 37$ strong evidence against H_0 *vis-à-vis* the alternative hypothesis H_1 stating that the draws were made *without* replacement? How about $X = 50$?

(c) Consider another alternative, H_2, stating that the draws were with replacement, but that only 25 of the 100 balls are black. Is the observation $X = 37$ strong evidence against H_0 *vis-à-vis* H_2? [$\Pr_2(X = 37) = 0.002$, $\Pr_2(X \leq 37) = 0.997$.]

3.2 Verify that for n observations i.i.d. $N(\theta, \sigma^2)$, with σ^2 known, the 1/8 likelihood interval for θ is $\bar{x} \pm 2.039\sigma/n^{1/2}$ and that this is a 95.9% confidence interval. Find the 1/32 likelihood interval and its confidence coefficient.

3.3 One form of reasoning that is sometimes used in efforts to give confidence intervals an evidential interpretation is as follows: The fact that a confidence interval procedure rarely results in the true value's being excluded implies that when a value is excluded, there is strong evidence that it is not the true one. Use the example in Exercise 2.4 to show that this reasoning is faulty.

CHAPTER 4

Paradigms for statistics

> In fact, as a matter of principle, the infrequency with which, in particular circumstances, decisive evidence is obtained, should not be confused with the force, or cogency, of such evidence.
>
> (Fisher, 1959, p. 93)

4.1 Introduction

We have examined three different approaches to formulating and solving the problem of testing statistical hypotheses. The Neyman–Pearson approach was the subject of Chapter 2, while two others, both called significance tests, appeared in Chapter 3 (p-value procedures and rejection trials). These correspond to three different views of the objective of a scientific study or experiment and of the role of statistical methods. They represent three distinct paradigms for statistics. In the next section we consider the reasons for the popularity of these three approaches, as well as their common shortcoming. We propose an alternative, likelihood-based, paradigm that will generate and support a more satisfactory theory and a more useful body of methods for representing, interpreting, and communicating statistical evidence in scientific studies. We describe the new paradigm in section 4.3 and begin to examine its use in sections 4.4–4.6.

4.2 Three paradigms

One important point of agreement among the three approaches to hypothesis testing is their view of the kinds of study or experiment to which they apply. They apply when it is reasonable to treat the basic observations as realizations of random variables, and when there is uncertainty about the probability distributions of those random variables. They all use a probability model for the experiment, a model which represents the observations $x_1, x_2, \ldots, x_n$ as realizations of random variables $X_1, X_2, \ldots, X_n$ whose probability distribution is the object of study.

A very simple example of such a study or experiment is the physician's diagnostic test in Chapter 1. The observation, x, indicating whether the test is positive or negative, is modelled as a realization of a random variable X whose probability distribution is given by the first row of Table 1.1 if the patient has the condition and by the second row if he does not. The test is given in order to produce evidence about whether he has the condition, which is equivalent under the statistical model to evidence about which of these two probability distributions is generating X. Of course, most experiments involve observations x that take many more than the two possible values, $+1$ and -1, in this example, and most probability models involve many more than two possible distributions for X. For example, the number of successes X in ten Bernoulli trials has eleven possible values, 0, 1, ..., 10, and every value of the success probability, $0 < \theta < 1$, determines a different probability distribution for X.

The problem whose formulation and solution by Neyman and Pearson has served as a paradigm for modern statistics is one like our diagnostic test example, in which the probability model for the experiment consists of only two probability distributions, corresponding to two simple statistical hypotheses. Neyman and Pearson viewed the experiment as a procedure for choosing between the two hypotheses, a procedure in which X is observed and its value determines which hypothesis is chosen. They argued that such 'hypothesis-testing' procedures should be evaluated in terms of their error probabilities, and they showed how to find, among those procedures whose Type I error probability does not exceed a specific bound, α, the one whose Type II error probability is smallest. That is, they found, among all hypothesis-testing procedures whose size is less than or equal to α, the one that is most powerful. The Neyman–Pearson theorem about most powerful test procedures is a cornerstone of statistical theory today. More generally, as we described in Chapter 2, the Neyman–Pearson formulation and solution of the problem of testing two simple hypotheses have served as a prototype for other statistical problems, such as testing composite hypotheses, estimation, and setting confidence limits.

Why has their result been so influential? What makes it so attractive? For one thing, their finding of the optimal test procedure made choosing a test an objective process, driven by the mathematics of size and power, rather than by the experimenter's subjective judgements about which test is most appropriate. But what is most valuable and compelling about the Neyman–Pearson view is a more fundamental contribution to the objectivity of experiments. The experiments to which their theory applies consist of making observations that, because of such factors as biological variation and measurement error, are unpredictable. Furthermore, after the observations have been made, they are found to be compatible with many hypotheses, no one being definitely proved or disproved. In such experiments there is an essential element of uncertainty which implies that the observations can be misleading; they can lead to rejection of a hypothesis that is true, to a confidence interval that misses the true value, or to an estimate with a large error.

The Neyman–Pearson formulation makes these aspects of the experiment explicit and gives the experimenter two remarkable advantages: it provides precise, objective measures of the probability of results that are misleading or otherwise unsatisfactory; and, even more important, it gives methods for controlling those probabilities. By choosing to test at the 0.05 level the experimenter controls the probability of a Type I error at that value. And by choosing the sample size, he controls the probability of a Type II error – he can set β at any positive value and, by choosing a sufficiently large sample (as in formula (2.1), for example), guarantee that the Type II error probability does not exceed β. It is these advantages, objective measure and control of the frequency of errors, that have earned statistical hypothesis testing an important place in science. Likewise, it is the provision of objective measure and control of their probabilistic properties (or 'performance characteristics', in Neyman's language) that has made other statistical procedures, such as those for estimation and for setting confidence intervals, into important scientific tools. By choosing the 95% confidence level the experimenter fixes the probability that his interval will miss the target value (at 0.05), and by choosing the sample size he controls the width of the interval.

But scientific experiments have a fundamental purpose that is not expressed in the Neyman–Pearson model. That purpose is to produce evidence about the phenomena being investigated. And a key role of statistics is to provide objective methods for representing, interpreting, and quantifying that evidence. We saw in Chapter 2 that the statistical methods derived from the Neyman–Pearson decision-making paradigm are not intended to meet this need.

In contrast to Neyman–Pearson hypothesis tests, significance tests do have 'evidence' as an explicit part of their terminology and rationale (Chapter 3). They begin with the same probability model as Neyman–Pearson theory (sometimes simplified so that there is only one hypothesis), but they view the purpose of the experiment differently: they view the experiment as a procedure for generating evidence against a hypothesis.

Significance tests of the type that we are calling rejection trials measure the evidence in terms of the significance level α that is chosen, with small α signifying strong evidence against the hypothesis. An important part of designing a rejection trial is choosing its significance level and the corresponding rejection region, R, whose probability is α or less if the hypothesis being tested is true. The experiment consists of observing X and noting whether it falls in R. If it does, it is said to be strong enough evidence against the hypothesis to justify rejection at significance level α. Here α plays two distinct roles: it is both the measure of strength of the evidence and the probability of obtaining misleading evidence (of at least that strength). In the latter role it enables the experimenter to measure and control the probability of a misleading result – generating evidence strong enough to justify rejecting a hypothesis that is actually true.

Significance tests of the other type, p-value procedures, associate with the observed value of X a number, the p-value, that is supposed to measure the strength of the evidence against the hypothesis – the smaller the p-value the stronger the evidence. Thus when the hypothesis is actually true, observations giving a small p-value represent misleading evidence. The probability of such misleading observations is controlled by the way p-values are constructed: if the hypothesis is true then the probability of obtaining evidence against it of strength α or greater (a p-value less than or equal to α) cannot exceed α. If the hypothesis is true, we will observe a p-value less than or equal to 0.05 no more than 5% of the time. Again the same probability has two distinct roles: the p-value is both the measure of strength of the evidence and the probability of obtaining misleading evidence.

As we saw in Chapter 3, the problem with significance tests is in their use of the significance level or p-value in the first of these two roles: to measure the strength of evidence. This use is based on the invalid law of improbability. The result is a methodology for interpreting data as evidence that is fundamentally wrong.

For a theory to support the use of current statistical methodology (hypothesis testing, confidence intervals, etc.) for representing and interpreting data as evidence we now have three unsatisfactory options, derived from the paradigms of Neyman and Pearson and the two varieties of significance testing. The first carefully excludes any concept of statistical evidence, while the other two try to build on the erroneous concept expressed by the law of improbability. Most textbooks of statistical theory embrace the Neyman–Pearson paradigm wholeheartedly. They concentrate on the pleasant mathematics involved in finding optimal procedures for hypothesis testing, estimation, etc., and ignore altogether the problem of interpreting data as evidence. A conspicuous exception is Cox and Hinkley's (1974) *Theoretical Statistics*, in which the authors observe that 'many if not most statistical analyses have as their immediate objective the summarization of evidence, the answering of the question "What can reasonably be learned from these data?" '. They present a theory shaped mainly by the Fisherian significance-test viewpoint, but appear to express concern for the discrepancies between that theory and the needs of science in warning that 'it is very important to ensure that the idealization which is inevitable in formulating theory is not allowed to mislead us in practical work' (Cox and Hinkley, 1974, p. 5).

The absence of an adequate theory would not be of pressing importance if our current statistical methods were satisfactory. In that case we would merely need a better rationale for what we do. But much of what we do is wrong. In Chapters 2 and 3 we illustrated the problems, using simple artificial examples for the most part. However, the problems appear also in important applications, as we saw in the case of the standard methods for determining sample sizes (section 2.4). Other important examples abound. For instance, the p-value's dependence on the sample space, illustrated in section 3.4 by the example with the misplaced code-book, leads to a methodology for analyzing data from sequential clinical trials that is widely recognized as unsatisfactory for representing observations as evidence (Cornfield, 1966; Armitage, 1975; Dupont, 1983).

The outlines of a more satisfactory theory have been around for some time. They were sketched by Fisher (1925), with major refinements from Barnard (1949), Birnbaum (1962), and Edwards (1972). We develop this theory in the next section in the form of a replacement for the Neyman–Pearson and significance-testing paradigms. It employs the same probability models as those theories do, so that, like them, it is non-Bayesian; but unlike them, it is compatible with Bayesian theories of decision-making and revision of opinion. (Bayesian statistics is examined briefly in Chapter 8.)

We will formulate the purpose of a scientific study as being the production of evidence – evidence that can be objectively interpreted and evaluated. This is not the only reason for doing a study, nor is it always the main reason. But it is usually an important reason, and when it is, the Neyman–Pearson and significance-testing formulations are inadequate, the former because it is concerned with a quite different purpose, and the latter because they lack an adequate means of representing and evaluating evidence.

4.3 An alternative paradigm

Statistical theory needs both an explicit quantitative concept of evidence and procedures for measuring and controlling the frequency with which experiments can produce weak or misleading evidence. In the absence of a direct expression for the evidence in current theories of hypothesis testing, it is inevitable that the quantities that are present, the error probabilities α and β, are routinely misinterpreted as measures of the strength of evidence, as is done when rejection of H_1 in favor of H_2 by a test with small α and β is taken to mean that we have strong evidence supporting H_2 over H_1, or when the demand for a small p-value is thought to be tantamount to a requirement for strong evidence.

The needed concept of evidence is embodied in the law of likelihood. And when evidence is measured by the likelihood ratio, as the law says it must be, we have the means both to specify the objectives accurately, in terms of the evidence that will be generated by the experiment, and to control the probabilities of unsatisfactory results. The potential for control was demonstrated in Chapter 1, where we pointed out a universal bound on the probability of misleading evidence, expression (1.2), and showed that we can make the probability of finding strong evidence supporting a true hypothesis over a false one as large as we like by making enough independent observations.

We begin with precisely the same model as the Neyman–Pearson theory: we have two hypotheses, H_1 and H_2, and these hypotheses imply respective probability density (or mass) functions, f_1 and f_2, for a random variable X. To obtain evidence about H_1 *vis-à-vis* H_2, we will perform an experiment – we will observe $X_1, \ldots, X_n$ i.i.d. as X. The result will be a data set, that is, a vector $(x_1, \ldots, x_n)$ of observations. We will interpret the data according to the law of likelihood – the ratio $L_2/L_1 = \prod_1^n f_2(x_i)/f_1(x_i)$ measures the evidence for H_2 versus H_1, and the bigger this ratio, the stronger the evidence. (A ratio less than one means that the evidence favors H_1.)

The purpose of the experiment is to produce strong evidence one way or the other, strong evidence supporting H_2 over H_1 or vice versa. The experiment will be unsuccessful if it produces a set of observations that constitute only weak evidence with respect to this pair of hypotheses. But weak evidence is not the only type of unsatisfactory result that the experiment can produce. It will be even worse if the experiment produces strong evidence in support of H_2 when H_1 is true or vice versa – weak evidence is bad, but strong evidence that is misleading is much worse.

To understand the similarities and differences between the Neyman–Pearson and evidential formulations of this simple problem it is useful to state them in the same format.

> *Neyman–Pearson formulation*: The experiment is a procedure for *choosing between* H_1 and H_2. It can result in an error, which can take either of two forms: choosing H_2 when H_1 is true (Type I), or choosing H_1 when H_2 is true (Type II). We want to be pretty sure (respective probabilities at least $1-\alpha$, $1-\beta$) that we will not commit an error of either type.

> *Evidential formulation*: The experiment is a procedure for *generating evidence about* H_1 *vis-à-vis* H_2. It can produce an unsatisfactory result, which can take either of two forms: strong evidence supporting the false hypothesis over the true one (misleading evidence), or weak evidence. We want to be pretty sure (respective probabilities at least $1-M$, $1-W$) that the experiment will not produce an unsatisfactory result of either type.

Strong evidence is represented by a likelihood ratio of at least some threshold value k. The observations are strong evidence in favor of H_2 if $L_2/L_1 \geq k$, strong evidence in favor of H_1 if $L_2/L_1 \leq 1/k$, and weak evidence if $1/k < L_2/L_1 < k$. What value of k is appropriate? The question is analogous to the one that is asked in problems of hypothesis testing: 'What significance level is appropriate?' Although no one choice will work for all problems, we can establish useful guidelines and conventions. The 5% significance level has been widely accepted as identifying evidence of sufficient strength to be of general scientific interest. Publication decisions made by researchers and by editors of scientific journals often depend critically on whether the results are 'statistically significant at the 5% level'. In the same way, a likelihood ratio of 8, for example, the ratio produced by observing three successive white balls in our canonical urn scheme, might come to be accepted as identifying evidence that is strong enough to be of general scientific interest under a wide range of conditions. When 'quite strong' evidence is sought, conventionally expressed by the choice of the

1% significance level, we might choose a threshold $k = 32$, corresponding to five successive white balls.

Of course, different values of k, say k_1 and k_2, might be appropriate for characterizing 'sufficiently strong' evidence in favor of H_2, $L_2/L_1 \geq k_2$, and 'sufficiently strong' evidence in favor of H_1, $L_2/L_1 \leq 1/k_1$. For example, if H_1 is widely believed, the experimenter might be satisfied with moderate evidence ($k_1 = 8$) in favor of H_1, but might judge that results supporting H_2 over H_1 will have little impact unless they represent evidence that is 'quite strong indeed', so that $k_2 = 64$ might be used in planning the experiment.

Having chosen k_1 and k_2, we can calculate the probability of failing to obtain sufficiently strong evidence, $\Pr(1/k_1 < L_2/L_1 < k_2)$ under H_1 and under H_2, and we can see how large n must be to ensure that these probabilities are acceptably small. That there will always be such an n is guaranteed by the fact, noted in Chapter 1, that both of the probabilities of finding strong evidence, $\Pr_2(L_2/L_1 \geq k_2)$ and $\Pr_1(L_2/L_1 \leq 1/k_1)$, approach one as n grows.

4.4 Probabilities of weak and misleading evidence: normal distribution mean

Consider for example the $N(\theta, \sigma^2)$ distribution with H_1: $\theta = \theta_1$ and H_2: $\theta = \theta_1 + \delta$ ($\delta > 0$, σ^2 fixed) that we used to illustrate the Neyman–Pearson sample-size calculations in Chapter 2. In this case the likelihood ratio L_2/L_1 equals $\exp\{[\bar{x} - (\theta_1 + \delta/2)]n\delta/\sigma^2\}$. Thus the observations will give an intermediate likelihood ratio $1/k_1 < L_2/L_1 < k_2$ (which means that the experiment will have failed to produce sufficiently strong evidence) whenever the observed sample mean $\bar{x}$ falls in the interval $(\bar{x}_{\mathrm{L}}, \bar{x}_{\mathrm{U}})$, where $\bar{x}_{\mathrm{L}} = \theta_1 + \delta/2 - \sigma^2 \log(k_1)/n\delta$ and $\bar{x}_{\mathrm{U}} = \theta_1 + \delta/2 + \sigma^2 \log(k_2)/n\delta$. If we denote the probabilities of this event, weak evidence, under H_1 and H_2 by W_1 and W_2 respectively, it is straightforward to calculate these values: defining the function $W(k_1, k_2)$ by

$$W(k_1, k_2) = \Phi\left(\frac{\sqrt{n}\delta}{2\sigma} + \frac{\sigma \log k_1}{\sqrt{n}\delta}\right) - \Phi\left(\frac{\sqrt{n}\delta}{2\sigma} - \frac{\sigma \log k_2}{\sqrt{n}\delta}\right),$$

we find

$$W_1 = \Pr_1(\bar{x}_{\mathrm{L}} < \bar{X} < \bar{x}_{\mathrm{U}}) = W(k_2, k_1),$$

$$W_2 = \Pr_2(\bar{x}_{\mathrm{L}} < \bar{X} < \bar{x}_{\mathrm{U}}) = W(k_1, k_2).$$

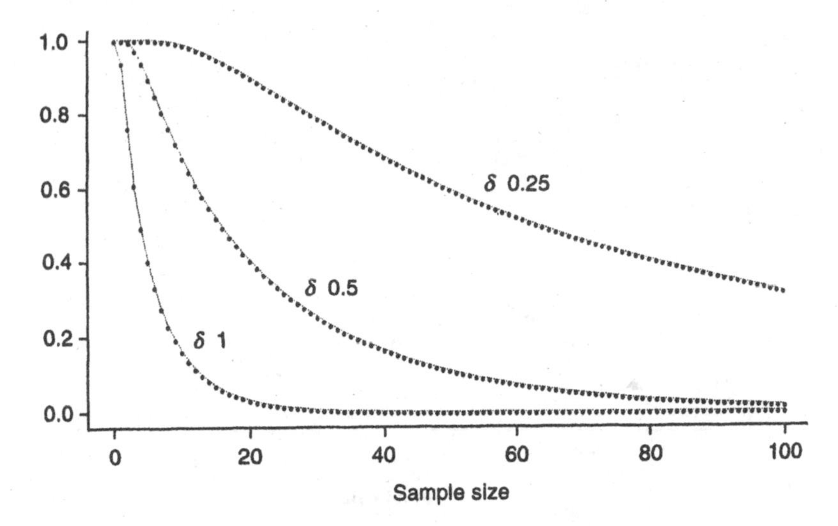

Figure 4.1 *Probability of weak evidence for normal mean:* $\frac{1}{8} < L_2/L_1 < 8$. *Hypothesized means differ by* δ *standard deviations.*

Clearly both W_1 and $W_2 \to 0$ as $n \to \infty$, so that we can set the risk of failure to provide sufficiently strong evidence at any value $0 < W < 1$ and find an n large enough to ensure that $\max(W_1, W_2) \leq W$.

When equal values are chosen for k_1 and k_2, the two probabilities of obtaining weak evidence, W_1 and W_2, are equal. For $k_1 = k_2 = 8$, Figure 4.1 shows how this common probability, $W(8, 8)$, decreases with increasing n when the two hypothesized means differ by 0.25, 0.5, and 1.0 standard deviations. To bring the probability of weak evidence down to 0.05, about 20 observations are required when the two hypothesized means differ by one standard deviation; about 60 are needed when the hypotheses differ by one-half of a standard deviation, and well over 100 are needed when the difference is only one-fourth of a standard deviation.

But what of the more serious undesirable outcome, strong evidence that is misleading? If H_1 is true this will happen whenever we observe a sample whose mean is greater than $\bar{x}_U$. Thus it will happen with probability $M_1 = M(k_2)$, where

$$M(k) = \Phi\left(-\frac{\sqrt{n}\delta}{2\sigma} - \frac{\sigma \log k}{\sqrt{n}\delta}\right).$$

If H_2 is true it will happen whenever the sample mean is less than $\bar{x}_L$, and the probability of this is $M_2 = M(k_1)$. Recall that the result

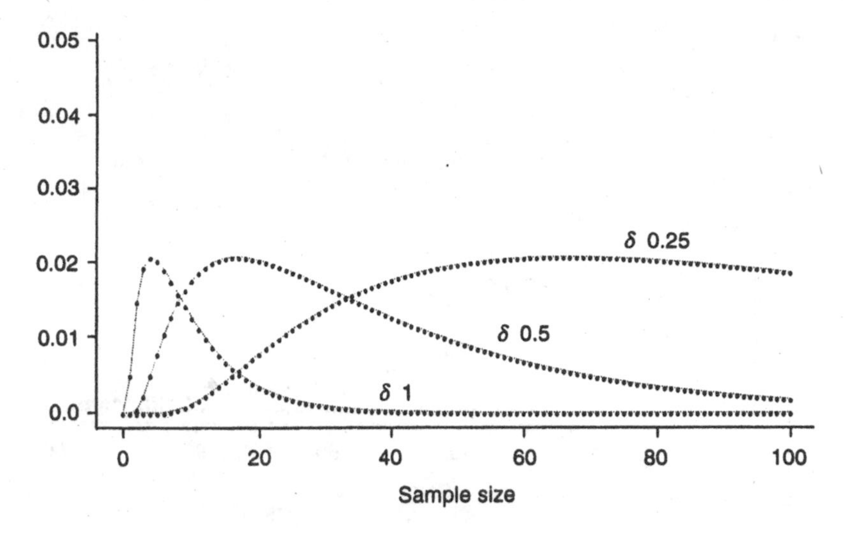

Figure 4.2 *Probability of misleading evidence for normal mean:* $\Pr_1(L_2/L_1 \geq 8)$.

proved in Exercise 1.5 shows that $M(k)$ can be no larger than $\Phi(-\sqrt{2\log k})$, for any choice of n. Thus for $k = 8$ and 32, $M(k)$ cannot exceed 0.021 and 0.004, respectively.

For any $k > 1$ and $\delta/\sigma \neq 0$, the behavior of the probability of misleading evidence, $M(k)$, follows the same pattern. It starts at zero when $n = 0$, rises steadily with increasing n until $n = (2\log k)(\sigma/\delta)^2$, where it reaches its maximum possible value, $\Phi(-\sqrt{2\log k})$, then decreases steadily thereafter. This is illustrated in Figure 4.2, which shows, for $k = 8$, how the probability of misleading evidence $M_1 = M_2 = M(8)$ varies with the sample size n, again when the two hypothesized means differ by 0.25, 0.5, and 1.0 standard deviations. Note that the upper limit on the vertical axis is only 0.05, and that for each curve the maximum value is $\Phi(-\sqrt{2\log 8}) = 0.021$.

From Figure 4.1 we see that when $\delta/\sigma = 0.25$ and the sample size is small (e.g. $n < 10$), the probability of weak evidence, W, is nearly one; that is, there is little chance of finding even moderately strong evidence ($k \geq 8$) in favor of either hypothesis when their means differ by only one-fourth of a standard deviation. Of course, this also means (Figure 4.2) that there is very little chance of misleading strong evidence. This is because we will have strong evidence for H_2 only if $\bar{X} \geq x_U$; when $n = 10$, for example, the probability of this under H_2 is $W_2 = 1 - \Phi(2.235) = 0.01$, and under H_1 it is even smaller, $M_1 = 1 - \Phi(3.026) = 0.001$.

The probability of misleading evidence, M_1, is in one sense analogous to the probability α that a Neyman–Pearson test procedure will reject H_1 when it is true, because rejecting H_1 is often taken (incorrectly) to signify that strong evidence against H_1 has been observed. To pursue the comparison, look at the sum $M_2 + W_2$, the probability, when H_2 is true, of either misleading strong evidence or weak evidence. This is the probability of failing to find strong evidence supporting H_2 when H_2 is actually true; it is analogous to the 'Type II error' probability β, the probability of failing to reject H_1 in favor of H_2 when H_2 is true. But the new paradigm recognizes that the probability of failing to find strong evidence for H_2 versus H_1 when H_2 is true consists of two very different components: the probability M_2 of finding misleading evidence in favor of H_1; and the probability W_2 of finding only weak evidence.

There is an important empirical distinction between misleading and weak evidence – when the experiment is finished, we know whether it produced weak evidence. That is, the event, weak evidence, whose respective probabilities under the two hypotheses are W_1 and W_2, will be seen to have occurred or not. When we obtain strong evidence, we cannot know if it is misleading or not, but when we obtain weak evidence we know it.

To emphasize the distinction between the error probabilities, (α, β), and the probabilities, $(M_1, M_2 + W_2)$ of misleading evidence in favor of H_2 and of failure to find strong evidence for H_2 when it is true, we graph these two quantities against sample size n in Figure 4.3. Here we chose the standard deviation $\sigma = 1$, with the hypothesized means differing by $\delta = 1/2$. We chose $k_1 = k_2 = 8$ to represent 'fairly strong evidence' and $\alpha = 0.05$. Specifically, the functions shown are:

$$\alpha = 0.05,$$

$$M_1 = \Pr_1(L_2/L_1 \geq 8) = \Phi(-\sqrt{n}/4 - 2(\log 8)/\sqrt{n}),$$

$$\beta = \Pr_2(\text{do not reject } H_1) = \Phi(1.645 - \sqrt{n}/2),$$

$$M_2 + W_2 = \Pr_2(L_2/L_1 < 8) = \Phi(-\sqrt{n}/4 + 2(\log 8)/\sqrt{n}).$$

Here α is the probability of *choosing* H_2 when H_1 is true, while M_1 is the probability of *finding fairly strong evidence in favor of* H_2 when H_1 is true; β is the probability of *failing to choose* H_2 when H_2 is true, while $M_2 + W_2$ is the probability of *failing to find fairly strong evidence in favor of* H_2 when H_2 is true.

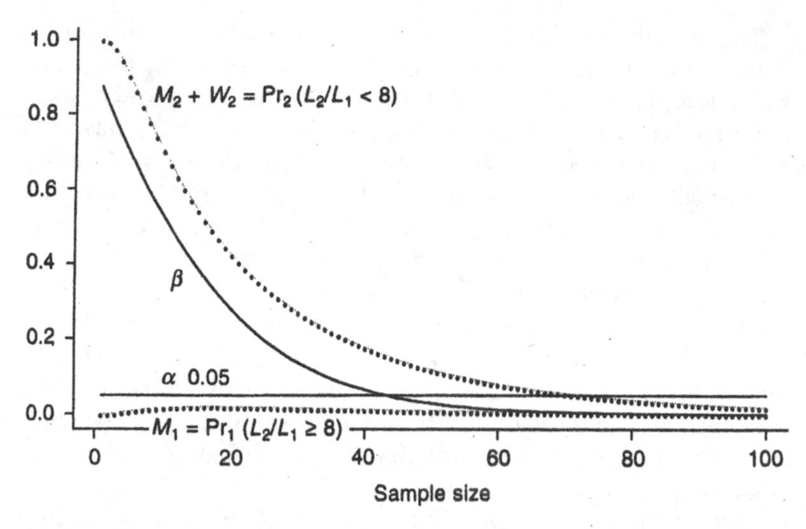

Figure 4.3 *Probabilities of undesirable results for normal mean.*

Under the new paradigm we will make no errors. The result of the experiment is not a decision, not an act of will, but a body of data, and we will interpret the data correctly. If $\boldsymbol{X} = \boldsymbol{x}$ is observed we will correctly interpret this as evidence supporting H_2 over H_1 by the factor $L_2/L_1 = \exp\{[\bar{x} - (\theta_1 + \delta/2)]n\delta/\sigma^2\}$. If this number falls between $1/k_1$ and k_2 we will be disappointed, for the evidence is not as strong as we had sought. If it is greater than k_2 or less than $1/k_1$, we will be satisfied to have found evidence (supporting H_2 over H_1 or vice versa) strong enough to meet our objectives. If the observed sample mean is slightly larger than $\theta_1 + \delta/2 + \sigma^2(\log 8)/n\delta$, then we have fairly strong evidence supporting H_2 over H_1. We have no way of knowing if that evidence is misleading. We do know that our interpretation of it is correct – it is indeed fairly strong evidence for H_2 *vis-à-vis* H_1. We also know that the process that generated it cannot often produce misleading strong evidence.

4.5 Understanding the likelihood paradigm

The new evidence-generating paradigm employs precisely the same probability model that the hypothesis-testing paradigm of Neyman and Pearson does. This model differs from the one used in the significance-testing paradigms only in that, like the Neyman–Pearson model, it must include an explicit alternative hypothesis. Whether the probabilities in the model represent physical tendencies, relative frequencies in some hypothetical sequence of repetitions, or subjective uncertainty has no bearing on the choice between paradigms. Whatever the probabilities mean, it is the same for each. In this respect, none is any more or less objective, frequentist, or non-Bayesian than another.

The critical difference between the new evidence-generating paradigm and the significance-testing paradigms is that the new one distinguishes between the strength of the evidence, which is measured by the likelihood ratio, and the probabilities that evidence of at least some specified strength will be generated under various conditions. (See the quote at the beginning of this chapter.) These probabilities – for example, the ones we denoted by M_1, M_2, W_1, and W_2 in the above discussion – can be important for planning a study and for deciding whether an ongoing study is worth continuing. But after a study is done they have nothing to do with the interpretation of a given body of observations as evidence. The old and new paradigms are entirely different in this respect, because in the old ones, the interpretation of a given body of observations as evidence is made in terms of the error probabilities. The new paradigm requires changing mental habits that have been established by decades of erroneously equating *p*-values with strength of evidence.

Consider, for example, three experiments similar to the diagnostic test for disease in section 1.2. In experiment 1 an observation is made on a random variable X_1, whose respective distributions under the two hypotheses about the patient's condition (H_1: disease present, and H_2: disease absent) are shown in the first box in Table 4.1. The observation $X_1 = +1$ is strong evidence for H_1 versus H_2 (likelihood ratio $94/2 = 47$), while $X_1 = -1$ is fairly strong evidence for H_2 (likelihood ratio $98/6 = 16.3$). Since these are the only two

Table 4.1 *Probability models for three experiments*

	Experiment 1		Experiment 2			Experiment 3	
	+1	−1	+1	0	−1	+1	−1
H_1	0.94	0.06	0.47	0.50	0.03	0.47	0.53
H_2	0.02	0.98	0.01	0.50	0.49	0.01	0.99

Table 4.2 *Probabilities (M_1, M_2) of strong evidence that is misleading ($LR \geq 8$) and probabilities (W_1, W_2) of weak evidence ($1/8 < LR < 8$) for H_1: disease present versus H_2: disease absent*

	M_1	M_2	W_1	W_2
Experiment 1	0.06	0.02	0	0
Experiment 2	0.03	0.01	0.50	0.50
Experiment 3	0	0.01	0.53	0.99

possible observations, experiment 1 will never produce weak evidence ($1/8 < L_2/L_1 < 8$), so that both W_1 and W_2 are zero. If H_1 is true then $X_1 = -1$ is misleading evidence in favor of H_2, and the probability of this is $M_1 = 0.06$. Likewise when H_2 is true $X_1 = +1$ is misleading strong evidence in favor of H_1, so that $M_2 = 0.02$.

Now consider two other tests or experiments that might be done. These consist of observing random variables X_2 or X_3, whose probability distributions under H_1 and under H_2 are also shown in Table 4.1.

Table 4.2 shows the probabilities of weak or misleading evidence in each of the three experiments when $k_1 = k_2 = 8$ is used to define strong evidence. Experiment 1 is a very good one – as we noted above, the probability that it will produce only weak evidence is zero; and it has a small probability, $M_1 = 0.06$, of producing misleading evidence when H_1 is true and a smaller probability, $M_2 = 0.02$, when H_2 is true. Experiment 2 has even smaller probabilities of misleading evidence, but half of the time it will produce $X_2 = 0$, which is utterly worthless evidence ($L_1/L_2 = 1$). Experiment 3 is worse – because $X_3 = +1$ is strong evidence for H_1 and $X_3 = -1$ is very weak evidence, experiment 3 can never produce strong evidence in favor of H_2 ($M_2 = 0.01$ and $W_2 = 0.99$, so $M_2 + W_2 = 1$). And when H_1 is true this experiment will fail to produce strong evidence more than half of the time ($W_1 = 0.53$).

These three experiments might all employ the same diagnostic test but different procedures for deciding when to perform the test or for recording the result. Experiment 2 might represent the procedure of a physician who asks for authorization to perform experiment 1, which is granted half of the time. When he fails to get the authorization, he gets the worthless observation, $X_2 = 0$. Experiment 3 might consist of performing experiment 2, but recording only whether $X_2 = +1$ was observed or not.

When any one of these three experiments has been performed and $X = +1$ has been observed, we have quite strong evidence in favor of H_1 versus H_2 ($LR = 47$). For interpreting this observation we need not know which of the three experiments produced it. The interpretation is correct regardless of whether it was experiment 1, 2, or 3. The likelihood function is the same two-point function in each case, and the likelihood principle guarantees that the evidence that we have is no stronger, no more reliable, no more significant if it arose from experiment 1 than if from one of the other two. The inferiority of experiment 3 (with respect to the probabilities of weak or misleading evidence) in no way weakens or contaminates the evidential meaning of the observation $X_3 = 1$ – the evidence supporting H_1 over H_2 is precisely the same as that in $X_1 = 1$ and that in $X_2 = 1$. If it makes sense to speak of the probabilities of H_1 and of H_2, then the ratio of the probability of H_1 to that of H_2 is increased by the same factor, 47, in all three cases.

The three experiments are procedures for generating evidence about H_1 *vis-à-vis* H_2. The probabilities in Table 4.2 are important for assessing what can be expected from the procedures; they properly measure the confidence we can have that a given procedure will produce strong evidence and the risk that it will produce misleading evidence. But they have no bearing on the reliability or strength of the evidence after it is produced.

4.6 Evidence about a probability: planning a clinical trial and interpreting the results

The previous section used a simple aritificial example to emphasize the distinction that is critical in the new paradigm, the distinction between planning a study and interpreting the observations that a study has produced. The probabilistic characteristics of the decision process that will guide the experiment and determine when it will stop, such as the probabilities that weak or misleading evidence will be obtained, are important at the planning stage. But for interpreting the results, likelihood ratios are the appropriate measures, and the probabilities of what might have been, what would have been done if the observations had been different, the unrealized intentions of the investigators, are all irrelevant.

Here is an example that is based on a real problem in clinical trials where the usual blend of Neyman–Pearson and Fisherian statistical ideas and methods produced quite a large mess (Bartlett *et al.*, 1985; Ware, 1989; Begg, 1990). In the early 1980s, the standard medical treatment for newborn babies with a certain critical health problem was unsatisfactory. Extensive clinical experience had shown that with what was then the standard treatment, the survival rate was only about 0.20. A radical new treatment, known as 'extracorporeal membrane oxygenation' (ECMO), had been developed, and preliminary studies were encouraging, suggesting that a survival rate of 0.80 or better might be achieved. Some were skeptical, however; they doubted that the survival rate with ECMO would actually be any better than the old rate, 0.20. With this background, let us imagine planning a study to learn more about the new treatment's success rate, θ. In particular, we want to generate strong evidence about the hypothesis H_2: $\theta = 0.80$ *vis-à-vis* H_1: $\theta = 0.20$. We will have at least fairly strong evidence for H_2 versus H_1 if $(0.8/0.2)^x(0.2/0.8)^{n-x} \geq 8$, or $x \geq (2n+3)/4$. Thus for $n = 7$, we will find at least fairly strong evidence supporting H_2 versus H_1 only if we observe five or more successes. If H_1 is true, then the probability of this (misleading evidence) is only $\Pr_1(X \geq 5) = 0.005$.

In this example the two hypothesized values of the ECMO success probability θ are equally far from $\frac{1}{2}$, so that for any choice of equal values $k_1 = k_2 = k$, the two probabilities of misleading evidence are equal, $M_1 = M_2$, as are the probabilities of weak evidence, $W_1 = W_2$. These common values, M and W, are shown in Table 4.3 for various choices of n when $k = 8$. There we see that if either H_1 or H_2 is true, then with only eight patients, we have a better

Table 4.3 *Probabilities of misleading (M) and weak (W) evidence for Bin(n, θ) distribution. H_1: θ = 0.2, H_2: θ = 0.8*

	$k = 8$		$k = 32$	
n	M	W	M	W
7	0.005	0.143	0.005	0.143
8	0.010	0.046	0.001	0.202
9	0.003	0.083	0.003	0.083
10	0.006	0.026	0.001	0.120
11	0.002	0.048	0.002	0.048
12	0.004	0.016	0.001	0.072
13	0.001	0.029	0.001	0.029
14	0.002	0.009	0.000	0.043
15	0.001	0.017	0.001	0.017
16	0.001	0.006	0.000	0.026
17	0.000	0.010	0.000	0.010

than 95% chance of producing at least fairly strong evidence supporting one over the other, and we have only a 1% chance of producing fairly strong evidence in favor of the wrong hypothesis.

Because there is controversy over whether the new treatment really is effective, we might judge that the 'fairly strong' evidence represented by $k = 8$ is inadequate and aim for $k = 32$ instead. Table 4.3 also shows the probabilities for this objective. Now eight patients are not enough; a sample size of eight exposes us to an unacceptably high risk that if θ really is 0.8, we will fail to produce evidence of the desired strength ($W = 0.202$). If 5% is an acceptable level for this risk, then 11 patients are sufficient. But if we want to be quite certain (with probability 0.99) that we will produce strong evidence (a likelihood ratio of at least 32) we need 17 patients in the study. And with that many subjects, we can be quite confident indeed (probability greater than 0.999) that our study will not produce misleading strong evidence. (Exercise 4.1 provides the opportunity to explore the patterns that appear in Table 4.3.) Graphs of the values in Table 4.3 are shown in Figures 4.4 and 4.5.

When the study is done we will interpret and report the results via the likelihood function for θ. The likelihood principle assures us that for interpreting the observations as evidence about the ECMO success rate we need the likelihood function and nothing more. Let us examine the evidence in some of the possible outcomes.

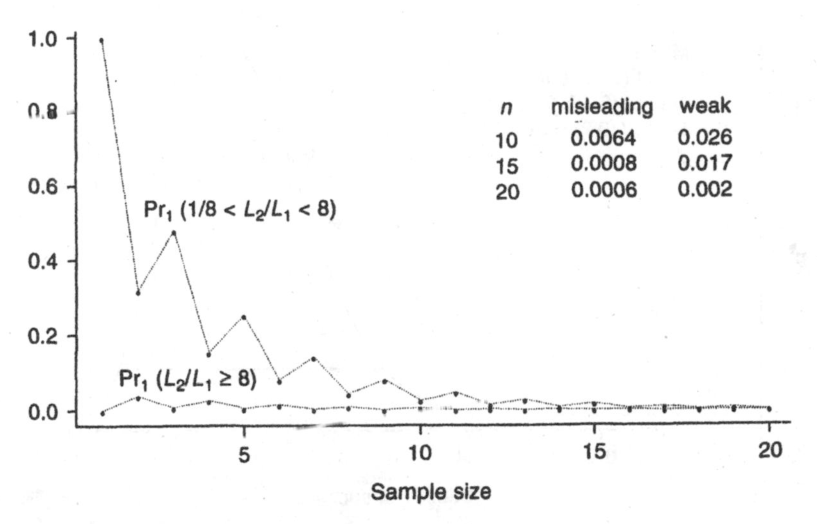

Figure 4.4 *Probabilities of weak and misleading evidence for binomial probability,* $k = 8$.

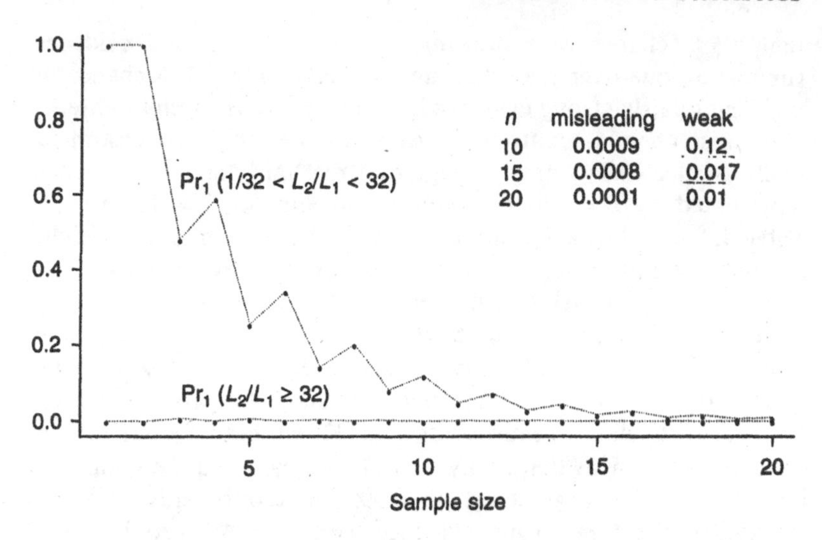

Figure 4.5 *Probabilities of weak and misleading evidence for binomial probability,* $k = 32$.

First, suppose that we perform 17 trials and observe 13 successes. This is pretty much what the proponents of the new treatment predicted, a success rate close to 80% ($13/17 = 0.76$). The likelihood function, proportional to $\theta^{13}(1-\theta)^4$, is shown in Figure 4.6. Since only ratios of values of the likelihood function are meaningful, the

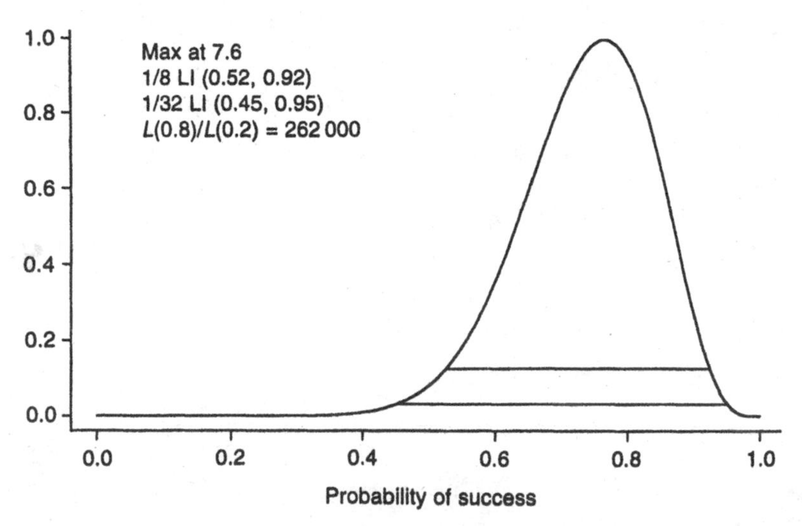

Figure 4.6 *Likelihood for probability of success: 13 successes observed in 17 trials.*

vertical scale is arbitrary, and in Figure 4.6 we have standardized it so that the maximum value is one. Examining certain features of the graph in Figure 4.6 corresponds to the activities of estimation, hypothesis testing, and setting confidence intervals. We note where the likelihood is greatest, the magnitude of the likelihood ratio at parameter values of special interest, and where the 1/8 and 1/32 likelihood intervals lie. These are shown in the upper left-hand corner of Figure 4.6 – the likelihood function is maximized at $13/17 = 0.76$, the 1/8 likelihood interval is from 0.52 to 0.92 and the ratio of the likelihood at $\theta = 0.8$ to the old value, $\theta = 0.2$, is 262 000. These features of the graph do not represent separate methods to be applied, each one representing the solution to a different problem of statistical decision-making. They are simply different aspects of the simple, comprehensive solution to the problem of representing the data as evidence, that is, the likelihood function.

These observations represent quite strong evidence indeed for $\theta = 0.8$ versus $\theta = 0.2$ – the likelihood ratio of 262 000 is evidence of the same strength as that in 18 consecutive white balls in the urn example of section 1.6. The 1/8 and 1/32 likelihood intervals are not confidence intervals, in general, but they truly represent what confidence intervals are often mistaken to represent, namely parameter values that the sample does not represent evidence against, that is, values that are 'consistent with the observations'. We can speak in this way, asserting that there is not strong evidence against a point inside the interval, without reference to an alternative value, because the statement is true for all alternatives. Every point inside the 1/8 interval is consistent with the observations in the strong sense that there is no other possible value of the parameter that is better supported by a factor as large as 8.

For points outside the likelihood intervals, the interpretation must be more careful. There is fairly strong evidence against a point just outside the 1/8 likelihood interval in the specific sense that there is *some* alternative value, namely the maximum likelihood estimate (MLE) that is better supported by a factor of at least 8. We saw in the example of drawing one card from a deck (section 1.7) that it is not correct to interpret such results as evidence against a single hypothesis. In the present example, the observations are pretty strong evidence supporting the MLE, $\theta = 0.76$, over the parameter value at the end of the 1/8 likelihood interval, $\theta = 0.52$, because $L(0.76)/L(0.52) = 8$. They are equally strong evidence supporting the latter value, $\theta = 0.52$, over the value $\theta = 0.42$

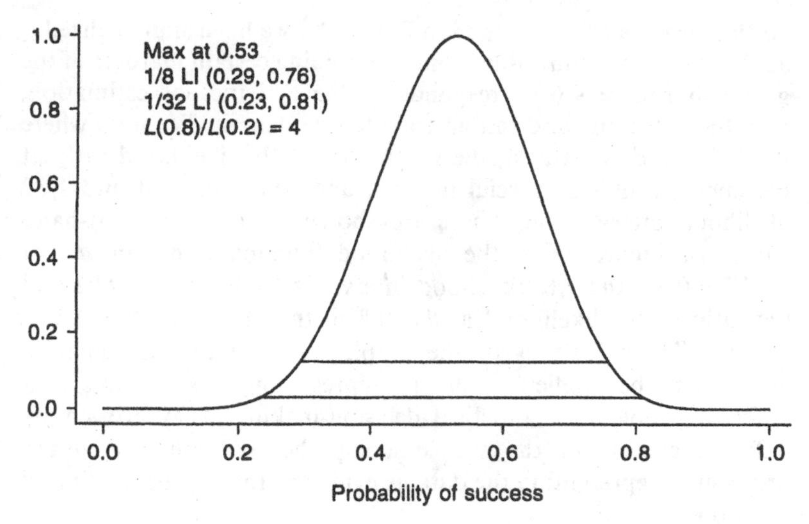

Figure 4.7 *Likelihood for probability of success: nine successes observed in 17 trials.*

where the standardized likelihood function equals 1/64, because $L(0.52)/L(0.42) = 8$. They are not 'evidence against $\theta = 0.52$'.

Next suppose that different results are observed. Suppose that the new treatment is successful in only nine of the 17 patients. In this case the likelihood function is shown in Figure 4.7. These observations are only weak evidence for $\theta = 0.8$ over $\theta = 0.2$ ($LR = 4$), but rather strong evidence for $\theta = 0.5$ versus $\theta = 0.2$ ($LR = 88.8$), and moderately strong evidence for $\theta = 0.5$ versus $\theta = 0.8$ ($LR = 22.2$). That is, we have rather strong evidence supporting success rates around 0.5 over the rate 0.2 associated with the old treatment, and fairly strong evidence for the intermediate rates versus the rate 0.8 that we were hoping to achieve. This evidence is unaffected by the fact that when we planned the study we were primarily interested in the two values $\theta = 0.2$ and $\theta = 0.8$, and gave little thought to other values. The likelihood function represents the evidence about all values of θ; for every pair of values, $L(\theta_2)/L(\theta_1)$ measures the evidence in our data for θ_2 versus θ_1. No special interpretation or penalty is attached to this ratio according to whether these values were or were not given special roles in the planning of the study. The probabilities, M and W, that were important when we planned the study are now entirely irrelevant – they have no role at all in interpreting our observations as evidence about the success rate.

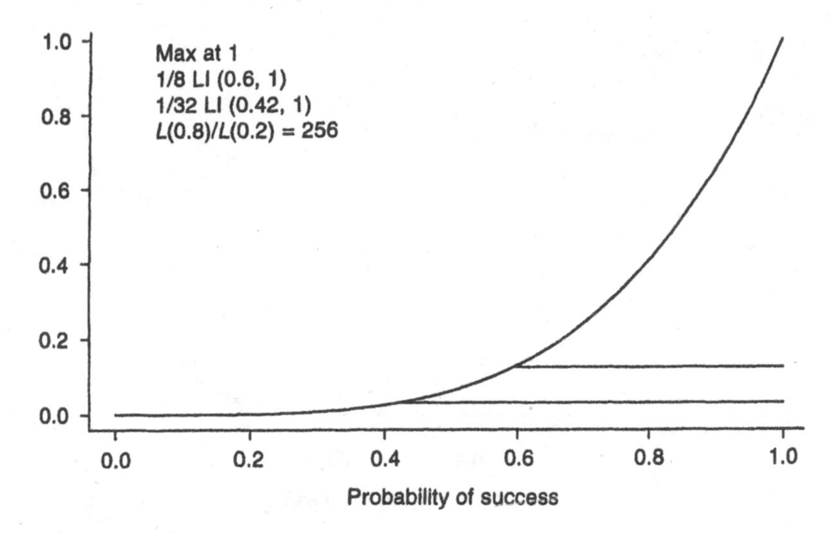

Figure 4.8 *Likelihood for probability of success: four successes observed in four trials.*

Since patients eligible for this study arrive one by one and we usually learn of the outcome for one study patient before the next one arrives, we can observe and analyze the results as we go along. If the treatment is successful in all of the first four patients, then we have a likelihood function proportional to θ^4, shown in Figure 4.8. These four observations are quite strong evidence for $\theta = 0.8$ over $\theta = 0.2$ ($L(0.8)/L(0.2) = 256$, evidence of the same strength as eight consecutive white balls in the canonical urn scheme in section 1.6). Even for a skeptic who does not take seriously the suggestion that ECMO might produce a survival rate as high as 0.8, but who considers $\theta = 0.6$ to be a plausible value, the evidence is strong ($L(0.6)/L(0.2) = 81$). We might well decide to stop the study at this point.

On the other hand, if the value of the success rate for the old therapy, which we have been taking to be $\theta = 0.2$, is not well documented and widely accepted, we might decide to continue with the study. If $\theta = 0.4$ is a plausible value for the old success probability, then our evidence for $\theta = 0.8$ versus that value is 16, which is less than our target of 32. And if the old treatment's success rate could well be as great at $\theta = 0.5$ in patients like these, then our evidence for $\theta = 0.8$ versus that value, $L(0.8)/L(0.5) = 6.6$, is not as strong as that of three white balls in the urn scheme.

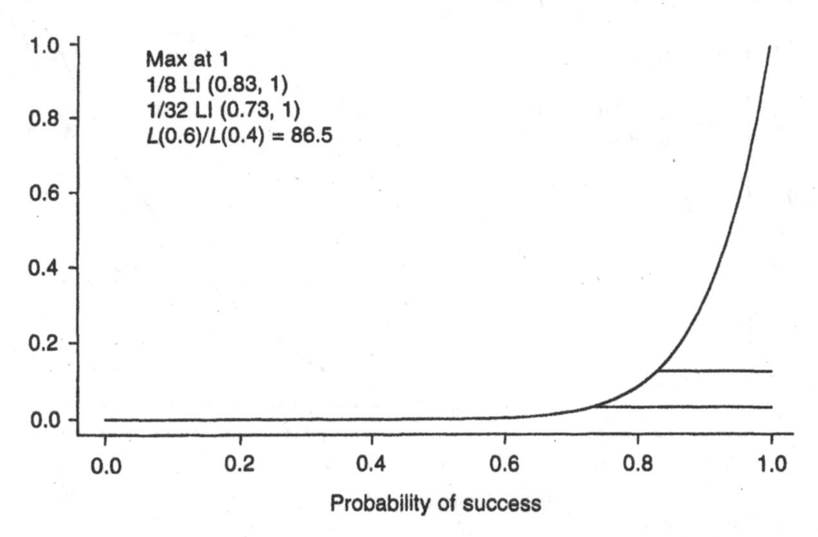

Figure 4.9 *Likelihood for probability of success: 11 successes observed in 11 trials.*

In the study that was actually performed, 11 patients were given the new treatment (ECMO), and all survived (Bartlett *et al.*, 1985). The evidence about the new treatment's success probability is represented by the likelihood function, $L(\theta) \propto \theta^{11}$, shown in Figure 4.9. It overwhelmingly supports $\theta = 0.8$ over $\theta = 0.2$. The likelihood ratio, $L(0.8)/L(0.2) = 4^{11}$ (greater than 4 million), is the same as that based on 22 successive white balls in the urn scheme. Now the evidence for $\theta = 0.8$ versus $\theta = 0.5$ is strong ($L(0.8)/L(0.5) = 176$). Even a skeptic who insists that the old rate is twice the value that the researchers' experience suggested, or $\theta = 0.4$, and who is unwilling to take seriously a success rate for the new therapy greater than $\theta = 0.6$, must concede that the observations constitute strong evidence for the superiority of the new therapy ($L(0.6)/L(0.4) = (1.5)^{11} = 86.5$).

The Bartlett *et al.* study produced strong evidence about the success rate for the new treatment. The proper interpretation of that evidence is made by applying the law of likelihood to the function shown in Figure 4.9, and it is not affected by whether the researchers fixed the number of observations at 11 in advance, or they originally planned to make 17 observations but stopped after 11 because the results were better than expected. Indeed, the interpretation is independent of whatever rule they might have used to decide when to make more observations, and when to stop. Such rules affect the probabilities, M and W, associated with the study, but they do not affect the strength or reliability of the evidence that the study produced.

Not all patients in the Bartlett *et al.* study received the new treatment. The study was actually a randomized clinical trial that used a rather complicated 'randomized play-the-winner' rule for deciding when to stop, and, when the rule called for continuation, for deciding whether the next patient would receive ECMO or the old therapy, designated as 'conventional medical theory' (CMT). In addition to the 11 subjects in the study who received ECMO and survived, there was one who was randomly assigned to receive CMT and died. Thus the overall evidence that the survival rate is greater under the new treatment was even stronger than we suggested in the above discussion.

Current beliefs about statistical inference in clinical trials are dominated by extreme skepticism regarding the use of information not generated within the immediate study. This is why the researchers in our example felt compelled to use both treatments in their study despite their belief that the new one was superior. This is also why conventional analyses of their observations ignored the existing evidence about the dismal success rate of the old therapy, and considered only the study's internal observations of 'zero successes in one trial' under the old treatment in evaluating the evidence about the relative merits of the two therapies.

But even after agreeing to confine their attention to the 12 observations made within the study, statisticians using standard frequentist methods were unable to reach a consensus about the strength of the Bartlett *et al.* study's evidence in favor of the new therapy. For example, they produced p-values and cited α-levels (for testing the hypothesis of no difference between treatments) ranging from 0.038 and 0.051, through 0.28 and 0.50, up to 0.62 (Begg, 1990). This was caused specifically by uncertainties about aspects of the study that the law of likelihood shows to be clearly irrelevant, namely, what rules the researchers actually followed in making decisions as the study progressed, and by how those rules should in principle affect the statistical analysis of the results. The likelihood function for the odds ratio in favor of survival under ECMO versus CMT is shown in Figure 4.10. (Figures 4.10 and 4.11 actually show *conditional* likelihoods for the odds ratio. These will be discussed in Chapter 7.) Survival rates of 0.8 for ECMO and 0.2 for the old therapy give an odds ratio of $\psi = 16$ and the evidence in favor of this value versus $\psi = 1$ is moderately strong: $L(16)/L(1) = 7.11$.

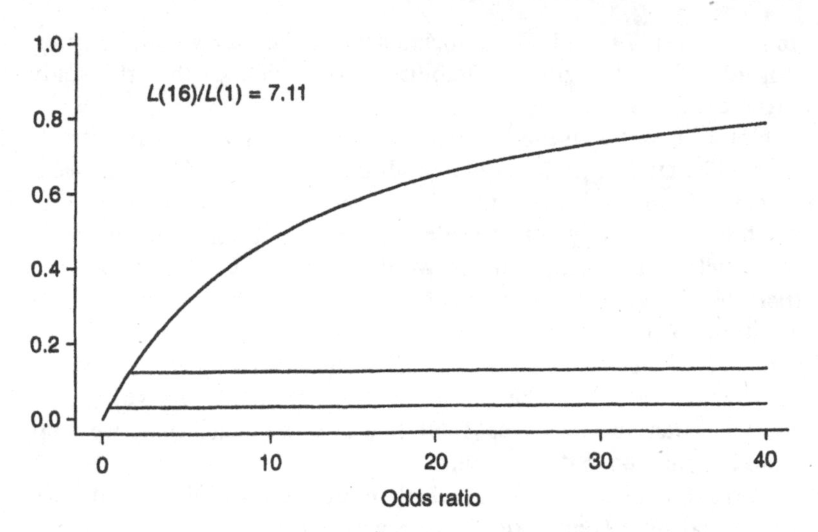

Figure 4.10 *Likelihood for ratio of odds on survival: ECMO versus CMT (Michigan study).*

Larger values of ψ are even better supported by these observations, and the lower end-point of the 1/8 likelihood interval is at $\psi = 1.59$.

The statistical confusion had important consequences. In a commentary on the Bartlett *et al.* study, Ware and Epstein (1985) stated that 'the type I, or false positive, error rate for this design is 0.5!' and concluded that 'Further randomized clinical trials using concurrent controls [i.e. patients randomized to the old therapy] ... remain necessary'. This led to the performance of yet another clinical trial in which patients were randomized to ECMO or CMT until there were four deaths in one treatment group. After 19 patients were randomized there were four deaths in the CMT group. At that point randomization ended, and the next 20 patients were assigned to ECMO. Results for all 39 patients are shown in Table 4.4 (O'Rourke *et al.*, 1989). This trial sparked debates about both the

Table 4.4 *Results of ECMO study II*

	Survival	
	+	−
ECMO	28	1
CMT	6	4

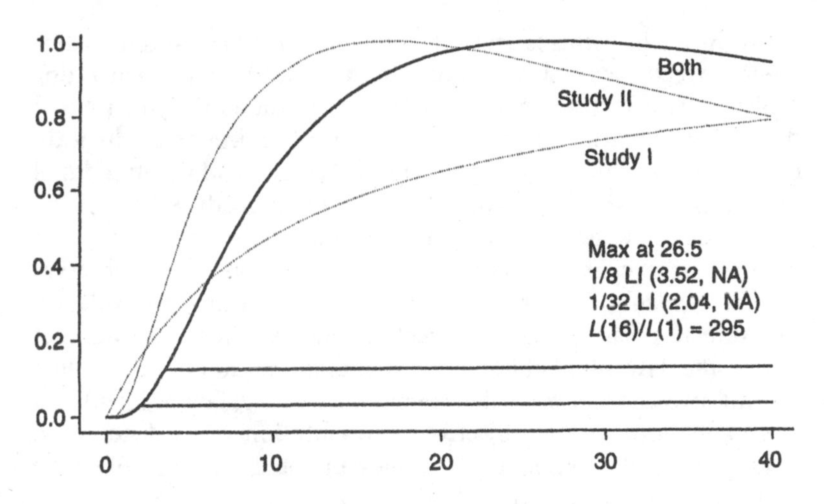

Figure 4.11 *Likelihood for ratio of odds on survival: ECMO versus CMT (two studies).*

statistical evaluation and the medical ethics of randomized clinical trials (Ware, 1989; Royall, 1991).

Figure 4.11 shows the evidence about the odds ratio in favor of survival under ECMO in each of the two studies, and in the two together. (For the second study above, the lower end-points of the 1/8 and 1/32 likelihood intervals are 2.00 and 1.12, respectively.) Unfortunately, the story does not end here. Some discussants of Ware's (1989) paper, again driven by the illogic of the old paradigms, continued to argue that the evidence against the null hypothesis (no treatment difference) was weak, and these arguments were cited in explaining why yet another randomized trial of ECMO was performed (United Kingdom Collaborative ECMO Trial Group, 1996). That large study randomized more than 90 babies to each therapy and resulted in 25 more infant deaths under CMT than under ECMO.

4.7 Summary

Statistics gives science the capacity to measure and control the uncertainty that is a prominent characteristic of many experiments and observational studies. This is true of both the Neyman–Pearson and Fisherian schools of statistics. The former is guided by a paradigm that represents an experiment as a procedure for choosing between two hypotheses and the latter by significance-testing paradigms, which view an experiment as a procedure for generating evidence with respect to one specified hypothesis and then either deciding whether it is strong enough to justify rejecting the hypothesis (rejection trials) or giving a numerical measure of the strength of the evidence against the hypothesis (p-value procedures).

Statistics, in either the Neyman–Pearson or Fisherian form, does not provide science with valid techniques for representing observations as evidence, or for measuring the strength of such evidence. But statistics has the key to meeting this requirement, and that key is the law of likelihood. A paradigm built on the explicit quantitative concept of evidence embodied in the law of likelihood can fulfill both needs: objective representation of evidence (and measurement of its strength); and measurement and control of the probabilities of weak and misleading evidence.

The probabilistic properties of procedures for generating and analyzing statistical data are important for planning experiments. But after an experiment is done, these properties are not appropriate for representing and interpreting the evidence that has been produced. Nevertheless, their use for that purpose is central to scientific applications of the statistical methods based on the Neyman–Pearson and significance-testing paradigms. The new paradigm restricts these probabilistic properties to their proper role (planning); it represents and interprets observed statistical evidence in terms of likelihood ratios, in accordance with the law of likelihood.

Exercise

4.1 Explain why the probabilities of weak and misleading evidence (W and M) in Table 4.3 for $n = 7$ are the same when $k = 8$ as when $k = 32$.

CHAPTER 5

Resolving the paradoxes from the old paradigms

5.1 Introduction

Statistics has many peculiarities and paradoxes that are the inevitable result of using methods derived from the Neyman–Pearson and/or significance-testing paradigms for purposes of representing and interpreting statistical data as evidence. In the light of the new likelihood paradigm we can understand the peculiarities and resolve the paradoxes. In this chapter we look at some conspicuous examples.

5.2 Why is a power of only 0.80 OK?

We begin with a mild peculiarity – why is it that the Type I error rate α is ordinarily required to be 0.05 or 0.01, but a Type II error rate as large as 0.20 is regularly adopted? This often occurs when the sample size for a clinical trial is being determined. In trials that compare a new treatment to an old one, the 'null' hypothesis usually states that the new treatment is not better than the old, while the alternative states that it is. The specific alternative value chosen might be suggested by pilot studies or uncontrolled trials that preceded the experiment that is now being planned, and the sample size is determined using a formula like (2.1) with $\alpha = 0.05$ and $\beta = 0.20$. Why is such a large value of β acceptable? Why the severe asymmetry in favor of α? Sometimes, of course, a Type I error would be much more costly than a Type II error would be (e.g. if the new treatment is much more expensive, or if it entails greater discomfort). But sometimes the opposite is true, and we never see studies proposed with $\alpha = 0.20$ and $\beta = 0.05$. No one is satisfied to report that 'the new treatment is statistically significantly better than the old ($p \leq 0.20$)'.

Often the sample-size calculation is first made with $\beta = \alpha = 0.05$. But in that case experimenters are usually quite disappointed to see what large values of n are required, especially in trials with binomial (success/failure) outcomes. They next set their sights a bit lower, with $\alpha = 0.05$ and $\beta = 0.10$, and find that n is still 'too large'. Finally they settle for $\alpha = 0.05$ and $\beta = 0.20$.

Why do they not adjust α and settle for $\alpha = 0.20$ and $\beta = 0.05$? Why is small α a non-negotiable demand, while small β is only a flexible desideratum? A large α would seem to be scientifically unacceptable, indicating a lack of rigor, while a large β is merely undesirable, an unfortunate but sometimes unavoidable consequence of the fact that observations are expensive or that subjects eligible for the trial are hard to find and recruit. We might have to live with a large β, but good science seems to demand that α be small.

What is happening is that the formal Neyman–Pearson machinery is being used, but it is being given a rejection-trial interpretation. The quantities α and β are not just the respective probabilities of choosing one hypothesis when the other is true; if they were, then calling the first hypothesis H_2 and the second H_1 would reverse the roles of α and β, and $\alpha = 0.20$, $\beta = 0.05$ would be just as satisfactory for the problem in its new formulation as $\alpha = 0.05$ and $\beta = 0.20$ were in the old one. The asymmetry arises because the quantity α is being used in the dual roles that it plays in rejection trials – it is both the probability of rejecting a hypothesis when that hypothesis is true and the measure of strength of the evidence needed to justify rejection. Good science demands small α because small α is supposed to mean strong evidence. On the other hand, the Type II error probability β is being interpreted simply as the probability of failing to find strong evidence against H_1 when the alternative H_2 is true – that is, it is being taken to mean what the sum $M_2 + W_2$ actually means. When observations are expensive or difficult to obtain we might indeed have to live with a large probability of failure to find strong evidence. In fact, when the expense or difficulty is extreme, we often decide not to do the experiment at all, thereby accepting values of $\alpha = 0$ and $\beta = M_2 + W_2 = 1$.

The reasoning of the experimenter, when expressed verbally in terms of strength of evidence, is quite correct. Its expression is impossible in terms of the concepts and calculations of Neyman–Pearson theory, and problematic in terms of those of rejection trials. This reasoning finds proper mathematical expression only under the new paradigm, in which a large likelihood ratio represents what a small α is supposed to convey, strong evidence, and $M_2 + W_2$ truly represents what β is supposed to, the probability of failing to find strong evidence for H_2 over H_1 when H_2 is true.

5.3 Peeking at data: repeated tests

Cornfield (1966) described a striking peculiarity:

> The following example will be recognized by statisticians with consulting experience as a simplified version of a very common situation. An experimenter, having made *n* observations in the expectation that they would permit the rejection of a particular hypothesis, at some predesignated significance level, say .05, finds that he has not quite attained this critical level. He still believes that the hypothesis is false and asks how many more observations would be required to have reasonable certainty of rejecting the hypothesis if the means observed after *n* observations are taken as the true values. He also makes it clear that had the original *n* observations permitted rejection he would simply have published his findings. Under these circumstances it is evident that there is no amount of additional observation, no matter how large, which would permit rejection at the .05 level. If the hypothesis being tested is true, there is a .05 chance of its having been rejected after the first round of observations. To this chance must be added the probability of rejecting after the second round, given failure to reject after the first, and this increases the total chance of erroneous rejection to above .05... Thus no amount of additional evidence can be collected which would provide evidence against the hypothesis equivalent to rejection at the $P = .05$ level ...

When the observed means correspond to a scientifically important effect, and particularly when the magnitude of that effect is about what the experimenter's previous experience led him to expect, the conscientious statistician takes care to explain that failure to reject the null hypothesis does not mean that the evidence supports it – it means simply that the evidence against it is not strong enough to justify rejection at the chosen significance level. He might further explain that if the sample were larger, then these exact same observed means would justify rejection. Everything would be fine if the sample were larger.

The experimenter who then proposes to enlarge the sample and asks how many more observations are needed is dumbfounded by the statistician's response – no additional observations, no matter how many and no matter how extreme, will permit him to reject the hypothesis at $p = 0.05$; no additional observations will make his results 'statistically significant at the 5% level'. Although this answer sounds nonsensical, it is, as Cornfield explained, correct. Choosing to operate at the 5% level means allowing only a 5% chance of erroneously rejecting the hypothesis, and the experimenter has already taken that chance. He spent his 5% when he tested after the first n observations.

If the answer is correct, then why does it seem nonsensical? Again it is because the p-value is being used in two different ways, and one of these is inappropriate. First, it correctly measures the probability of erroneously rejecting the hypothesis. But it also is supposed to measure the strength of the evidence against the hypothesis: $p > 0.05$ means that the evidence is not very strong. Clearly, by taking more observations the experimenter can obtain stronger evidence. And some future observations, together with those already made, might represent very strong evidence. Therefore it must be possible, for some future observations, to obtain a very small p-value, contrary to what the statistician says.

A proper formulation of the problem resolves the paradox. The purpose of the experiment is to produce evidence. The 'hypothesis tested' is of special interest, but we can speak sensibly about the evidence for or against this hypothesis only in relation to alternative hypotheses. Suppose the experimenter's previous experience leads him to believe that the size of the treatment effect might well be as great as δ, say, and that an effect this large would clearly be of clinical importance. Then he might properly stop the study after the first n observations, when he first looks at the data, if the likelihood ratio

$$L(\delta)/L(0) = \exp\{(n\delta/\sigma^2)(\bar{x}_n - \delta/2)\}$$

exceeds some critical value such as $k = 8$, delimiting results that represent fairly strong evidence for a treatment effect of δ versus no effect. That is, he might properly stop if $\bar{x}_n > \delta/2 + \sigma^2 \log(8)/n\delta$. It is this condition, not the condition $\bar{x}_n > 1.645\sigma/n^{1/2}$ (i.e. p-value less than 0.05), that indicates moderately strong evidence against the hypothesis of no difference. If this is not achieved after n observations, there is no good reason, scientific or statistical, why he cannot make more observations. And his having examined the results after the first n does not preclude his later finding that the combined sample represents strong evidence supporting a treatment effect of δ versus no effect. In fact, he will have found such evidence after making m additional observations if the likelihood ratio for the combined sample,

$$L(\delta)/L(0) = \exp\{(m+n)(\delta/\sigma^2)(\bar{x}_{m+n} - \delta/2)\},$$

exceeds $k = 8$, that is, if $\bar{x}_{m+n} > \delta/2 + \sigma^2 \log(8)/(m+n)\delta$. The probability of finding strong but misleading evidence for a treatment effect of δ versus zero at the first look is not 0.05, but $1 - \phi(n^{1/2}\delta/2\sigma + \sigma\log(8)/n^{1/2}\delta)$, which cannot exceed $\phi(-\sqrt{2\log 8}) = 0.021$ for any combination of n, δ, and σ. The experimenter can 'use up' his Type I error probability, α, but he cannot 'use up' his probability of finding strong evidence supporting H_2 over H_1. The only way that he can preclude the possibility of obtaining future observations that, together with those already made, will represent strong evidence, is to accept the conclusions provided by the old statistical paradigms and stop the study.

The phenomenon illustrated by Cornfield's example is a central problem in clinical trials methodology. It is widely acknowledged that trials must be closely monitored, so that if early results are strong evidence that one treatment is better, the trial can be stopped – assigning late-entering subjects to a treatment with strong evidence against it would be ethically unacceptable. But 'monitoring' means looking at the early results, making a series of analyses on the accumulating data. The problem is precisely the one in Cornfield's example, and research in statistical methods for clinical trials has focused on 'the amount of Type I error that one can "use or spend" at each analysis' (DeMets, 1987). The tension caused by erroneously using the Type I error rate to measure the strength of the evidence is revealed in observations such as 'for a trial which continues to the end with an impressive trend [in which the final treatment means differ by more than 1.96 "standard errors"] but does not exceed [the larger critical value required to hold the total probability of Type I error at 0.05] the inability to reject H_0 can be awkward, if not difficult, to explain' (DeMets, 1987). To overcome this awkwardness, it has been suggested that two different significance levels should be used: the sequential (or 'overall') significance level, that reflects the increased Type I error probability caused by multiple looks; and the fixed-sample-size (or 'nominal') significance level, that is supposed to measure the strength of the evidence (Armitage, 1975, section 2.5; Dupont, 1983). This *ad hoc* suggestion represents a major step in the right direction, by making explicit the critical distinction between controlling the frequency of Type I error and measuring the evidence in a given set of observations. The problem, of course, is that it attempts to use p-values in both roles.

5.4 Testing more than one hypothesis

The previous section concerned the issue of adjusting statistical procedures to take account of 'multiple looks', or multiple tests of a single hypothesis on accumulating data. Another statistical pill that scientists sometimes have trouble swallowing is the adjustment of significance levels to account for testing multiple hypotheses. An experimenter using a normal distribution model can declare that an observed sample mean, $\bar{x}$, is statistically significantly different from zero at the 5% level if $|\bar{x}| > 1.96\sigma/n^{1/2}$. That is, the sample mean is statistically significant if it differs from zero by at least 1.96 times its standard error. She can then observe another mean, $\bar{y}$, and claim significance if $\bar{y}$ differs from zero by 1.96 times its standard error. She can do this, that is, if she observes and tests the means $\bar{x}$ and $\bar{y}$ in separate experiments. But if she efficiently combines the two experiments, she is likely to be advised that she must adopt a more stringent critical value. Now, she is told, she can declare $\bar{x}$ to be statistically significant only if it differs from zero by at least 2.24 standard errors (and similarly for $\bar{y}$).

The explanation is that the 1.96-standard-error criterion is appropriate for a single test, but now she has a multiple testing problem. The 1.96-standard-error rule will produce a 'false positive' result, a claim of statistical significance for a sample mean whose expected value is really zero, with probability 0.05. If she uses this rule in an experiment where two independent means, $\bar{x}$ and $\bar{y}$, are tested, and if both of their expected values are really zero, then the probability that she will commit a Type I error, declaring at least one of them to be statistically significant, is $1 - (0.95)^2 = 0.0975$, or almost twice the single-test error rate, 0.05.

To protect against this 'inflation of the error rate' she must adjust her test criterion. A simple way to do this is to use the Bonferroni inequality, which states that for k events $E_1, \ldots, E_k$, the probability that at least one occurs is no greater than the sum of their individual probabilities, $\sum \Pr(E_i)$. Thus if the probability that a single test will produce a Type I error is α, then the probability that at least one of k tests will produce an error is no greater than $k\alpha$. We can ensure that the probability of one or more errors in k tests does not exceed 0.05 by doing each individual test at the $0.05/k$ significance level. This is where the factor 2.24 standard errors comes from in the above case of $k = 2$ means, $\bar{x}$ and $\bar{y}$. The value 2.24 is the single-test critical value for $\alpha = 0.05/2 = 0.025$.

A researcher who plans an experiment that will test six different hypotheses, each one stating that an independent normal mean has expected value zero, can control her probability of committing a Type I error at 0.05 by performing each test at the 0.05/6 level, declaring each of the six sample means significant only if it differs from zero by 2.64 standard errors. If she does not use a 'multiple-test correction' and declares significance whenever a difference exceeds 1.96 standard errors, the probability that she will commit at least one Type I error might be as large as $1 - (0.95)^6 = 0.26$.

Awareness of the problem of inflation of error rates caused by multiple testing has increased in recent years, and now it is common for reviewers of applications for research grants and referees of journal articles to insist on Bonferroni-type adjustments to *p*-values and significance levels. (Only the most sophisticated referees suggest that the corresponding adjustments are required for confidence coefficients, but the same reasoning applies there as well.)

Although multiple test adjustments have become popular, they are sometimes resisted. The experimenter who is primarily interested in studying the distribution of X, and who plans a sample survey to select subjects for observation, will ordinarily make observations on other random variables, Y, Z, etc., in the same survey. Because identifying, contacting, recruiting, and interviewing subjects is an expensive business, it would be wasteful to ask only one question. The study gathers extensive data about each subject (e.g. age, medical history, diet, attitudes, blood and urine samples) and attempts to answer multiple research questions.

To the researcher who conceived the study in the first place to obtain evidence about the distribution of X, who finds a sample mean that differs from zero by 2.5 standard errors, and who is now told that this observation is not statistically significant, something seems wrong. If he had not taken advantage of the opportunity to test hypotheses about those other variables, Y, Z, etc., saving those questions for another study, then these same observations would have justified his rejecting the hypothesis that $\mathrm{E}X = 0$. But now they do not. His sense that the interpretation of his evidence about the distribution of X should not be determined by his actions with respect to other variables is correct, but it finds no expression within the three hypothesis-testing paradigms. The problem again is that the significance level is being used in two roles, only one of which is valid. Whether or not the researcher performs tests of other hypotheses affects his overall probability of committing at least one Type I error, but it does not change his evidence about the distribution of X or how that evidence should be interpreted. In particular, it does not weaken his evidence about the hypothesis $\mathrm{E}X = 0$ *vis-à-vis* any alternative value.

This does not mean that the researcher is justified in reporting unadjusted *p*-values or in declaring statistical significance according to the single-test criterion. It simply shows once again that p-values and significance levels do not measure the strength of statistical evidence. The evidence about EX is represented by the likelihood function, which measures the relative support for pairs of values. As we noted in section 3.10, the hypothesis stating that $EX = 0$ is unquestionably false so that there is actually no chance of committing a Type I error, and the question 'How strong is the evidence in our observations against that hypothesis?' is ill-formulated, as we saw in Chapter 1. The evidence supporting a specific alternative value of EX *vis-à-vis* zero is measured by the likelihood ratio, and it is unaffected by the researcher's attitude or behavior with respect to other hypotheses or questions.

5.5 What is wrong with one-sided tests?

Another bit of statistical advice that causes scientific discomfort is the recommendation that hypothesis tests should ordinarily be two-sided – that is, that one-sided p-values should routinely be doubled. This rule is sometimes applied even when there is reason to expect a departure from the null hypothesis in one specific direction, such as in an epidemiologic study of the effect of exposure to some agent on the probability of developing a specific disease. If the incidence of disease among subjects who have been exposed to the agent exceeds the incidence among the unexposed by an amount D, then the one-sided p-value is $\Pr_0(D \geq d_{obs})$. Researchers are often advised that the observed difference, d_{obs}, can be declared significant at the 5% level only if this probability does not exceed 0.025. For example, in a recent book review in a statistical journal, we find:

> the authors suggest if a one-sided test is used and the 'other side' ends up statistically significant, then the finding must be ignored. It strains credulity to imagine such self-control, and I prefer coming down strongly against one-sided tests. (Louis, 1993)

As this quotation illustrates, opposition to one-sided tests draws on a combination of statistics and psychology. If, contrary to expectations, the observed difference had been negative, the researcher might have reported a statistically significant difference in favor of the putative causative agent, evidence that it actually protects against the disease. That is, he might have rejected the hypothesis of no effect if he had found difference in either direction. If this is true, it implies that he will reject the null hypothesis when it is actually true twice as often as the one-sided significance level indicates. His actual Type I error rate is twice the level of the one-sided test, so the value he reports must be twice the one-sided p-value.

A researcher might be forgiven if he finds this argument presumptuous and condescending. If he says that he expected the difference to be positive, and that he is therefore justified in performing a one-sided test, and if he claims that he would not have altered the test if he had observed a negative difference, who is the statistician-psychologist to contradict him? By what authority does the statistician deny that the researcher would have simply declared 'no significant increase' if the observed difference had been negative? Would he, despite his assurances, have changed his hypothesis and tested for a decrease if that had been suggested by the data? Who knows? What we do know is that this evidence about the effect of the exposure on the disease incidence does not depend for its proper interpretation on the answer to such questions.

This common situation illustrates that p-values are not the straightforward objective measures of evidence that they are widely believed to be. They are not determined by what the experimenter did and what he observed. They depend also on what he would have done if the observations had been different – not on what he says he would have done, or even on what he thinks he would have done, but on what he really would have done (which is, of course, unknowable).

5.6 Why not use the most powerful test?

Consider a clinical trial in which subjects are paired. A randomly selected member of each pair is given a new treatment, while the other is given the old standard treatment. The results for the two subjects are compared and scored as 1 or 0, indicating whether the subject given the new treatment fared better or not. When ten pairs have been observed, the data might be analyzed under a probability model that represents the observations as realizations of i.i.d. *Bernoulli*(θ) random variables, X_i, $i = 1, \ldots, 10$, where X_i is the score for the ith pair.

A standard statistical analysis for this problem tests the hypothesis that the new treatment is no better than the old, H_1: $\theta \leq 1/2$, versus the alternative that it is, H_2: $\theta > 1/2$. Subject to the conventional constraint that the Type I error probability must not exceed 0.05, the standard test rejects H_1 in favor of H_2 if $\sum x_i \geq 9$. This is the standard test despite the fact that there is another test that controls the Type I error rate at the desired 5% level, and that has greater power (smaller β) against every single value of $\theta > 1/2$. That is, there is another 5% level test of H_1 that is uniformly more powerful against H_2 than the standard one.

The existence of the more powerful test means that, for a given sample size the researcher can, by simply using a different statistical test procedure, reduce the probability of a Type II error at every $\theta > 1/2$. Why does the less powerful test remain the standard?

The reason is that there is something about the most powerful test that does not make sense. Like the standard test, it calls for rejection of H_1 whenever $\sum x \geq 9$. In addition, it calls for rejection sometimes, but not always, when $\sum x = 8$. Specifically, when $\sum x = 8$ the most powerful test procedure requires generating a *Uniform*$(0, 1)$ random variable U, then rejecting H_1 if $U < 0.895$. Now the probability that this test will reject H_1 equals $\Pr(\sum X \geq 9) + 0.895 \Pr(\sum X = 8)$, which equals 0.05 when $\theta = 1/2$, and is less than 0.05 for all $\theta < 1/2$. Thus this test has size $\alpha = 0.05$. It is obviously more powerful than the standard test, since it will reject H_1 whenever the standard test does, and sometimes when it does not. In fact, at the alternative value $\theta = 3/4$ the power is increased by more than 100%, from 0.24 to 0.50.

The increase in power is possible because the size of the standard test is only 0.011, which is less than the allowable rate of 0.05. Here the most powerful test is an example of a randomized test. It can be proved that its power against any alternative $\theta > 1/2$ is greater than that of any test that depends only on the number of successes and has size at most 0.05 (e.g. the standard test).

Using the randomized test means letting the rejection of H_1 in favor of H_2 depend not only on the relevant statistical evidence, $\sum x$, but also on the clearly irrelevant observation, u, the value of the uniform random variable U. It means that the observation of $\sum x = 8$ is judged to be sufficiently strong evidence against H_1 to justify its rejection at the 5% significance level when $u \leq 0.895$, but not when u is larger. This makes no sense, because the evidence about θ (and about H_1 and H_2) is obviously the same in both cases. That the evidence is indeed independent of u is shown by the fact that when $\sum x = 8$, the likelihood function for θ is $\theta^8(1 - \theta)^2$, and is quite independent of u.

Here again it is the dual role of the size, α, that creates the problem. As long as α and β are interpreted as error rates only, the randomized test, being more powerful, is the more desirable. Dissatisfaction with the randomized test arises when α is also employed to measure the strength of the evidence, as it ordinarily is in applications like the present example.

This example illustrates again the distinction between the Neyman–Pearson and Fisherian schools of statistics. Fisherians explicitly use α as a measure of strength of evidence. To explain what is wrong with the more powerful test, and why the less powerful one should be used instead, they invoke a rule based on the concept of conditionality (Cox and Hinkley, 1974). That rule says that when probabilistic quantities such as *p*-values, confidence coefficients, standard errors, etc. are used for interpreting statistical data as evidence, these quantities should not be influenced by averaging over the sample space of a random variable like U, whose probability distribution does not depend on any unknown parameters. Because the rationale for conditioning relates not to efficiency, but to a concept of evidential interpretation that is part of the Fisherian paradigm, but foreign to their own, Neyman–Pearson theorists have been unable to find a place for conditionality within their theory (Lehmann, 1993).

5.7 Must the significance level be predetermined? And is the strength of evidence limited by the researcher's expectations?

Suppose the researcher does not specify the value of α before the experiment, but waits until after the observations have been made. She calculates the value of the test statistic, and if it exceeds the 1% critical value, she claims the results are 'significant at the 1% level'. If the test statistic falls between the 5% and 1% critical values, she claims the results are 'significant at the 5% level'. Under both the Neyman–Pearson and rejection-trial paradigms, these conclusions both imply that the hypothesis is rejected. And under both paradigms the test procedure is improper. This is because, although sometimes the result will be reported as 'significant at the 1% level', H_1 will actually be rejected whenever the test statistic exceeds the 5% critical value. In such cases the report is misleading, indicating that the test procedure has a Type I error probability of only 0.01, when it is really 0.05.

In order to control the Type I error probability at a fixed level, the value of α must be chosen before the experiment is done, and the researcher must resist the temptation to quote a smaller value in those cases where, after the data are in, it is found that they would have allowed rejection if the smaller value had been chosen. For this reason, authors adopting either the Neyman–Pearson or the rejection-trial perspective criticize the use of *p*-value procedures. Thus the psychologist Bakan (1970) wrote:

> The practice of 'looking up' the *p* value for the [*t*-statistic], ... rather than looking up the *t* for a given *p* value, violates the inference model. The inference model is based on the presumption that one initially adopts a level of significance ...

In a more recent critique of statistical methods in psychology, Dar, Serlin, and Omer (1994) described 'inappropriate use of null hypothesis tests and *p* values' as one of 'three major problems in statistical practice', explaining that:

> The sample *p* value, in the context of null hypothesis testing, is involved in a binary decision – specifically, in determining whether the predetermined criterion of Type I error, the alpha level, has been surpassed. Researchers, however, often make much more of the *p* value, turning it into the oracle of truth. Thus, they may linger over the obtained *p* values, honoring small values with three asterisks and with expressions such as 'highly significant' ... [R]esearchers cannot [both] use the obtained *p* values descriptively and endow them with binary inferential significance. This attempt to have the cake and eat it too is exemplified by the practice of taking any obtained *p* value and rounding it upward, thus creating post hoc an impression of predetermined alpha level (e.g. a *p* value of .0027 is reported as $p < .005$), a practice that results in numerous pseudo-alpha levels in a single study.

The rejection-trial paradigm requires that the significance level be predetermined, and that it cannot be renegotiated after the data have been examined. But because the significance level is also used to measure the strength of the evidence, this means that we cannot find stronger evidence than we set out to find – if our preselected significance level is 0.05, supposedly indicating only 'fairly strong' evidence, then no observations can justify our claiming significance at the 1% level, which would indicate evidence that is considerably stronger. This, of course, makes no scientific sense, and it explains why hypothesis testing, as practiced and reported in scientific journals, continues to consist largely of *p*-value procedures, despite criticisms such as those of Bakan and of Dar, Serlin, and Omer.

Consider the example of section 5.6, the experiment in which ten independent *Bernoulli*(θ) observations are made. For comparing evidence for H_1: $\theta = 1/2$ *vis-à-vis* H_2: $\theta = 3/4$, if the experimenter specifies the 0.05 significance level, then he can claim 'statistical significance' if nine or ten successes are observed. (If he uses the randomized test, he can also claim significance sometimes when he sees only eight successes.) But his statistical analysis cannot distinguish between the fairly strong evidence in favor of H_2 over H_1 that is represented by nine successes and the stronger evidence represented by ten. Regardless of which result he observes, nine successes or ten, he can only report that his observations are 'statistically significant at $\alpha = 0.05$'. The new paradigm explains what is wrong – the evidence is measured, not by '$\alpha = 0.05$', but by the likelihood ratios, $(1.5)^9(0.5) = 19.2$ for nine successes, and $(1.5)^{10} = 57.7$ for ten.

5.8 Summary

Statistical practice based on the Neyman–Pearson and Fisherian paradigms is, in important respects, counterintuitive and even paradoxical. The problems are usually found to arise from attempts to use probabilities to measure evidence. The ease and clarity with which the new likelihood paradigm explains these problems is testimony to its power and validity.

Exercises

5.1 (Evidence in results of multiple Neyman–Pearson tests of a fixed pair of hypotheses.) Suppose that $x_1, \ldots, x_c$ are the results of c independent tests of H_0 versus H_1; $x_i = 0(1)$ means that the ith test calls for choosing $H_0(H_1)$. The tests are all done at size α and power $1 - \beta$. When the available data consist of the observations $x_1, \ldots, x_c$ only (not the full data sets leading to these test results), how strong is the evidence in favor of H_1 over H_0 when k of the tests choose H_1 ($\sum_{i=1}^{c} x_i = k$) and $c - k$ choose H_0? If $\alpha = 0.05$ and $1 - \beta = 0.80$, calculate and interpret the results for all possible values of k when $c = 1$, 2, 3 and 5.

5.2 (Evidence in results of Neyman–Pearson tests of multiple hypotheses.) Suppose that the available data consist of observations $x_1, \ldots, x_c$ representing the results of independent tests of c unrelated pairs of hypotheses, H_{0i} versus H_{1i}, $i = 1, \ldots, c$. The ith observation $x_i = 0(1)$ indicates that H_{0i} (H_{1i}) is chosen. All tests are done at the same size α, but at different powers, $1 - \beta_i$, $i = 1, \ldots, c$. Consider the case when only one of the tests, say the kth one, calls for rejection of the first hypothesis in favor of the second, that is, $x_k = 1$, and $x_i = 0$, $i \neq k$.

(a) Let H_1^k: H_{1k} and H_{0i}, $i \neq k$, be the hypothesis stating that only one of the alternatives is true, and it is the kth one; and let H_0: H_{0i}, $i = 1, \ldots, c$, be the hypothesis stating that none of the alternatives is true. Show that those observations are evidence supporting H_1^k over H_0 by the factor $(1 - \beta_k)/\alpha$. That is, the evidence in $x_1, \ldots, x_c$ in favor of H_1^k over H_0 has the same strength as the evidence in the single observation, $x_k = 1$, in favor of H_{1k} over H_{0k}.

(b) Show that for any $j \neq k$ these observations support H_0 over H_1^j: $\{H_{1j}$ and H_{0i}, $i \neq j\}$ by the factor $(1 - \alpha)/\beta_j$.

(c) Let H_1 be the hypothesis that exactly one of $\{H_1^j, j = 1, \ldots, c\}$ is true, and that, furthermore, the probability that H_1^j is the true one is $1/c$, $(\Pr(H_1^j|H_1) = 1/c, j = 1, \ldots, c)$. Show that the strength of the evidence in favor of H_1 over H_0 is $\bar{\beta}/(1 - \alpha) + [(1 - \alpha - \beta_k)/\alpha(1 - \alpha)]/c$, where $\bar{\beta} = \sum \beta_i/c$, which equals $(1 - \beta_1)/\alpha$ when $c = 1$, but which is approximately $\bar{\beta}/(1 - \alpha)$ when c is large. Show that when the tests are all performed at $\alpha = \alpha^*/c$ (as suggested by the usual Bonferroni inequality adjustment for multiple tests when α^* is the desired overall Type I error rate), the likelihood ratio in favor of H_1 over H_0 is slightly greater than that in favor of H_{1k} over H_{0k} at level α^* when the only observation is $x_k = 1$ (i.e. $(1 - \beta_k)/\alpha^*$), but smaller than $(1 - \beta_k)/(\alpha^* + 1)$.

CHAPTER 6

Looking at likelihoods

6.1 Introduction

Both the Neyman–Pearson and the Fisherian views of statistics recognize three major divisions – estimation, hypothesis testing, and confidence intervals. These divisions represent a fragmented, piecemeal approach to analyzing and interpreting statistical evidence. Each of the three in turn has its own complicated set of concepts and techniques that must be used expertly. For example, what it means when a hypothesis test does not call for rejection must be understood in terms of: size and power; the distinction between 'not rejecting' a hypothesis and 'accepting' it; the distinction between 'statistical significance' and practical (or scientific) significance; the way that sample size affects the meaning of *p*-values, etc. We have seen that such matters are complicated enough that experts cannot agree.

On the other hand, recognizing that likelihood functions are the proper means for representing statistical evidence simplifies and unifies statistical analysis. Likelihood functions quantify the strength of statistical evidence in a form that lends itself to visual as well as numerical representation. We can look at graphs of likelihood functions and literally see what the data say.

This chapter consists of a series of examples in which statistical evidence is represented and interpreted via likelihood functions. Most of these examples use data from textbooks, manuals, and review articles, so the reader can conveniently compare what the likelihood functions show to the results obtained by skillful application of standard statistical methods. And in most cases the usual mixture of *p*-values, point estimates, and confidence intervals produces a reasonably accurate impression of 'what the data say', that is, they are in reasonable agreement with the likelihood analysis presented here. One reason for this agreement is that likelihoods play an important role in current methods for finding point estimates and approximate standard errors to use in producing confidence intervals and p-values. But because this 'likelihood methodology' is driven by the old paradigms it is formally incompatible with the law of likelihood and violates the 'irrelevance of the sample space' (section 1.11). Thus different methods are needed for evidential interpretation of data, that is, methods for examining and presenting likelihood functions *per se*.

The following examples show some of what has been done by an inexperienced user of S-Plus computer software. In some cases we take only a superficial look at the evidence, while in others we probe more deeply. Examining evidence under alternative models, and comparing data with observations artificially generated by simple models, should be a part of most analyses, but only a limited sample of such activities can be presented here. Thus these are not offered as thorough or definitive evidential analyses of these data sets. They are intended simply to show how revealing it can be to look directly at likelihood functions in order to see what the data say. Some of these examples use probability models that contain nuisance parameters in addition to the parameters of immediate interest. The rationale and techniques used to eliminate the nuisance parameters will be examined in Chapter 7.

6.2 Evidence about hazard rates in two factories

Swan (1993) used results from an occupational health study (Table 6.1) to show how data on deaths, for known person-years at risk, can be analyzed with GLIM computer software. Under the usual model specifying constant hazard rates within each of eight age-by-factory cells, the likelihood function for the rate λ in a cell where D deaths and P person-years at risk are observed is proportional to $(\lambda P)^D e^{-\lambda P}$ (Berry, 1983). Figures 6.1(a) and 6.1(b) show the eight

Table 6.1 *Data on deaths for known person-years at risk*

	Factory			
	1		2	
Age range	Deaths	Person-years	Deaths	Person-years
50–59	7	4045	7	3701
60–69	27	3571	37	3702
70–79	30	1777	35	1818
80–89	8	381	9	350

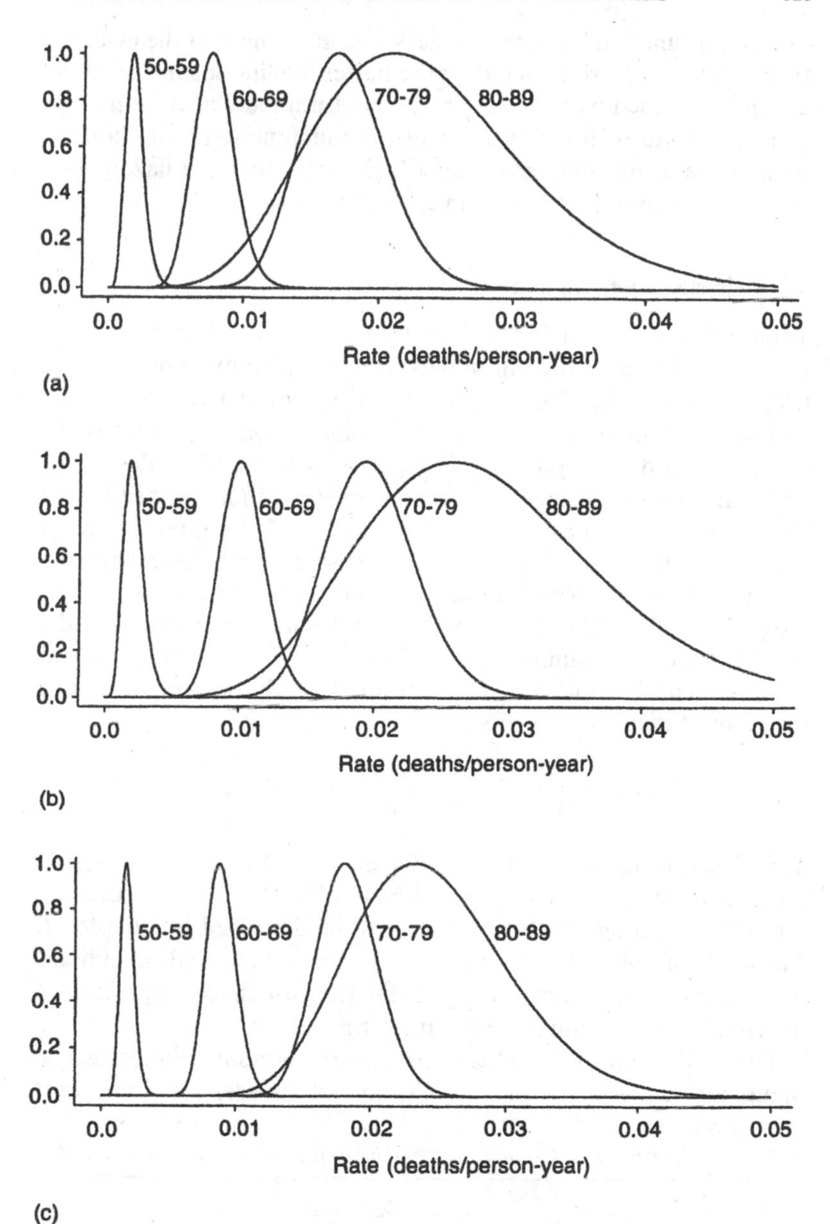

Figure 6.1 *Likelihoods for hazard rates, by age: (a) factory 1; (b) factory 2; (c) combined.*

likelihood functions. There is remarkable agreement in the evidence from the two factories, including the flatter likelihoods in the 80–89 age group caused by the smaller observed numbers of person-years at risk. Figure 6.1(c) shows the likelihood functions for the four parameters in the simpler model which states that the hazard rate in each age group is the same in both factories.

6.3 Evidence about an odds ratio

Table 6.2 contains data on the presence or absence of BCG vaccination scars among children in Malawi with and without leprosy (from Clayton and Hills, 1993, p. 175). We use a model that treats the number of children who have been vaccinated (as evidenced by the scar) in each disease group as independent binomial random variables, and focus on the evidence concerning a function of the two binomial probabilities: the ratio of the odds in favor of scars among those without leprosy to the odds among those with the disease. To obtain a likelihood function for the odds ratio, ψ, the most popular method is to use the conditional probability model in which the total number of vaccinated children, 7583, is treated as fixed. This conditional distribution depends only on the odds ratio, and the likelihood function is

$$L(\psi) \propto \left[\sum_{j=0}^{50} \binom{50}{j} \binom{13172}{7583-j} \psi^{j-22}\right]^{-1}.$$

This function is shown as the solid curve in Figure 6.2. Another method of deriving a single-parameter likelihood is to use the 'profile' likelihood function, which will be described in Chapter 7. The series of dots in the figure shows values of the profile likelihood for the odds ratio. Clearly it does not matter which technique is used for eliminating the nuisance parameter here.

The 'null' value of the odds ratio, 1, corresponding to the absence of any association between vaccination and disease, is an important

Table 6.2 *Presence of BCG scar among children with leprosy and without*

		BCG scar		
		+	−	
Leprosy	+	22	28	50
	−	7 561	5 611	13 172
		7 583	5 639	13 222

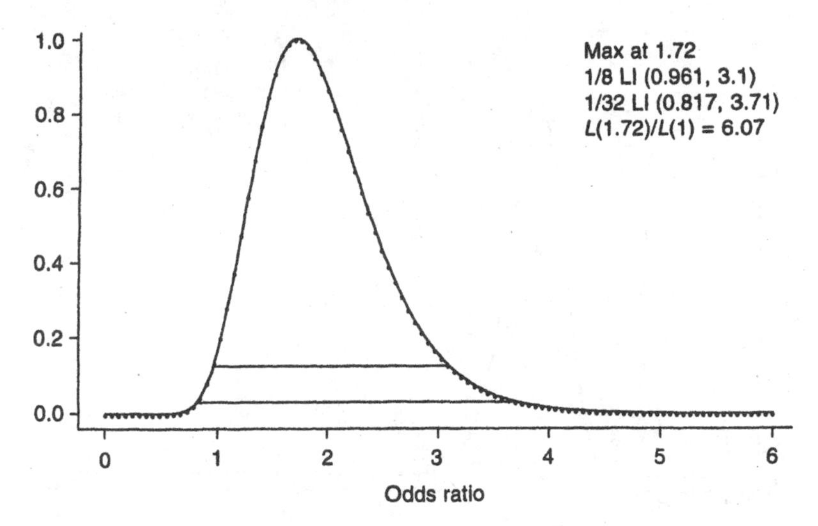

Figure 6.2 *Profile (dot) and conditional (line) likelihoods for ratio of odds: BCG scar in healthy subjects versus leprosy cases.*

reference point, and Figure 6.2 shows that there is modest evidence supporting odds ratios from about 1.5 to 2 over this value ($L(1.7)/L(1) = 6.1$, which has the same strength as 2.6 white balls in the urn scheme of section 1.6).

These data refer to children between 10 and 14 years of age. The incidence of leprosy increases with age, and Table 6.3 contains data for seven age groups, including the 10–14-year-olds. The (conditional) likelihood functions for the odds ratios in each of the seven age groups are shown in Figure 6.3 (dotted lines). Under a model

Table 6.3 *Presence of BCG scar among children with leprosy and without, by age*

	Without leprosy		With leprosy	
Age	Scar	No scar	Scar	No scar
0–4	11 719	7 593	1	1
5–9	10 184	7 143	14	11
10–14	7 561	5 611	22	28
15–19	8 117	2 208	28	16
20–24	5 588	2 438	19	20
25–29	1 625	4 356	11	36
30–34	1 234	5 245	6	47

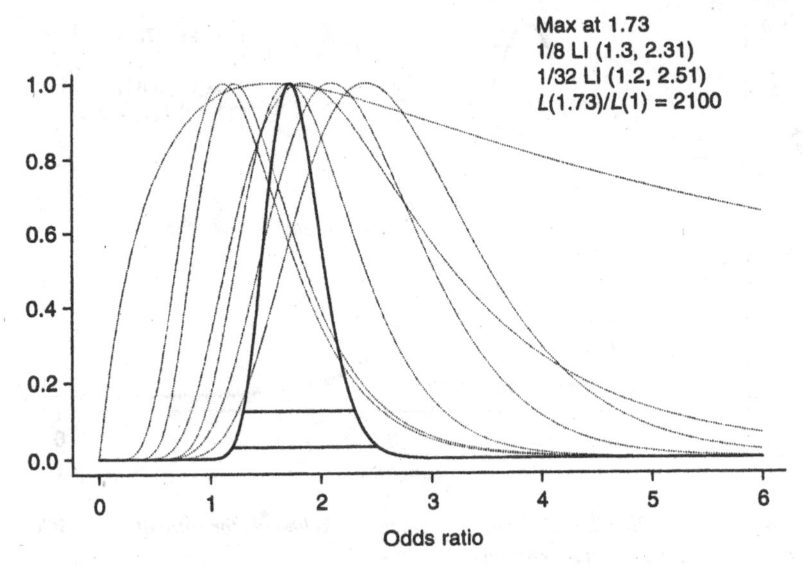

Figure 6.3 *Likelihoods for common odds ratio: BCG scar in healthy subjects versus leprosy cases.*

that specifies a common odds ratio ψ for all of these age groups, we have seven independent sets of evidence about ψ. In that case the likelihood for the entire set of seven age strata is proportional to the product of these seven functions. It appears as the solid curve in Figure 6.3, where the maximizing value of ψ and the 1/8 and 1/32 likelihood intervals also appear. The ratio of the likelihood at the MLE, $\psi = 1.73$, to the likelihood at the null value, $\psi = 1$, is over 2000. There is not strong evidence in favor of any value within the 1/8 likelihood interval (1.30, 2.31) versus any other, but the evidence in favor of any value within this interval versus $\psi = 1$ is quite strong ($LR > 2100/8 > 260$).

It is interesting to compare Figure 6.3 with the results of using the standard collection of statistical tools, which Clayton and Hills use this data set to illustrate. Under the model specifying a common odds ratio in all seven tables, Clayton and Hills evaluate a chi-square statistic with one degree of freedom for testing the hypothesis that $\psi = 1$ and find it to equal 15.2, giving a p-value of 1/20 000. Their calculations show that the standard Mantel–Haenszel estimator of the common odds ratio in these seven tables is 1.70, and that an approximate 95% confidence interval is (1.30, 2.23). Here the standard tools work well, in the sense that they give an overall assessment of the evidence that agrees well with the likelihood functions.

6.4 A standardized mortality ratio

Breslow and Day (1987, p. 72) use data from a long-term study of workers in a Montana copper smelter (Lee and Fraumeni, 1969) to illustrate various statistical methods for analyzing standardized mortality ratios (SMRs). The SMR for a specific cause of death is the ratio of observed to 'expected' deaths, O/E, where E is obtained by applying standard population rates to the exposure history of the study population. The denominator E is commonly described as the number of deaths expected in the study population if its rates were identical to those of the standard population.

Fifteen deaths from bladder cancer were observed in the study population, $O = 15$, while US population rates gave an expected number of only $E = 8.33$, for an SMR of $15/8.33 = 1.80$. The US population rates are given for five-year-by-five-year age-by-calendar-time cells, and under a model stating that the study population hazard rates in these cells are proportional to the US rates, the parameter of interest is the proportionality constant, ρ. This parameter is the ratio of the hazard rate in the study population to the rate in the standard population (the **hazard-rate ratio**) and the likelihood function is (Berry, 1983) $L(\rho) \propto (\rho E)^D e^{-\rho E}$.

The likelihood function for the bladder cancer hazard-rate ratio is shown in Figure 6.4. Again a key reference value for the ratio is one, indicating that the hazard rate in the study population is just the

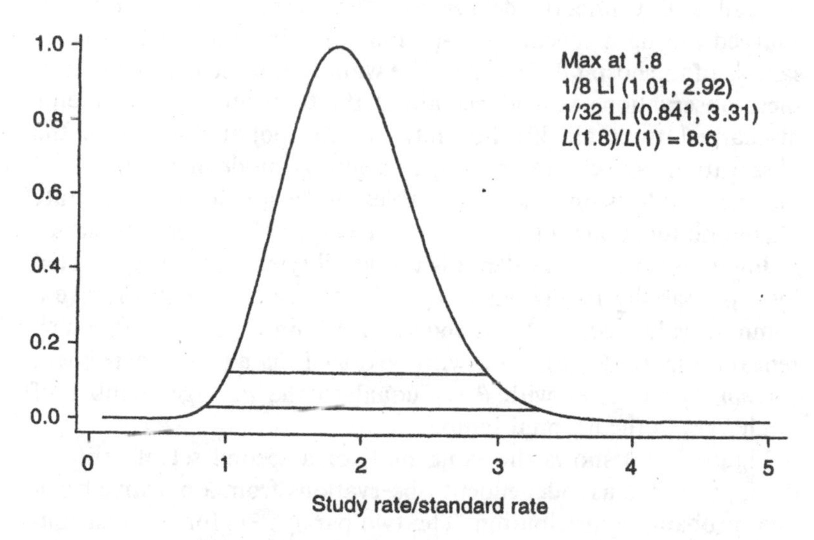

Figure 6.4 *Likelihood for hazard-rate ratio.*

same as that in the standard. These observations represent fairly strong evidence supporting a rate ratio of about two over a ratio of one: $L(1.8)/L(1) = 8.6$. That is, the evidence supporting the hypothesis that the study population's hazard rate is nearly twice the standard population's rate is fairly strong. They also represent fairly strong evidence supporting the hypothesis that the ratio of the study population's rate to the standard rate is only 1.8 over the hypothesis that the ratio is 3 or more.

Breslow and Day tried four different procedures for setting 95% confidence intervals for the hazard-rate ratio based on these observations and reported the results: (1.01, 2.97), (1.00, 2.98), (1.11, 2.97), (1.09, 2.99). Similarly, four procedures for calculating p-values to test the null hypothesis ($\rho = 1$) yielded 0.048, 0.049, 0.033, and 0.021, the last two being produced by the standard chi-square test with continuity correction and without. Again, there is good agreement between the evidential interpretation of these observations that is produced by the collection of techniques used by Breslow and Day and the interpretation represented by the likelihood function.

6.5 Evidence about a finite population total

A population of 393 short-stay hospitals has been used in several empirical studies of finite-population sampling theory and methods (Royall and Cumberland, 1981a). The number of patients discharged during a specific time-period was observed for a random sample of the hospitals ($n = 32$). We want to represent and interpret these observations as evidence about the total number of patients discharged from all 393 hospitals in this population. Since the observations are counts, we might begin by modelling them as 32 independent Poisson random variables, and examine the 32 separate likelihood functions, $L(\theta_i) \propto \theta_i^{y_i} e^{-\theta_i}$, $i = 1, \ldots, 32$. These are shown in Figure 6.5(a). This evidence is not at all typical of that generated by a probability model where the 32 Poisson distributions have a common value for θ. This is shown clearly in Figure 6.5(b), which repeats Figure 6.5(a), but with artificial data: 32 independent *Poisson*(θ) variables with θ set equal to the average number of discharges in the hospital sample.

Figure 6.5(c) shows the same plot for a second set of artificial data, generated as independent observations from a negative binomial probability distribution. The two parameters for this distribution are set so that the expected value is again equal to the discharge

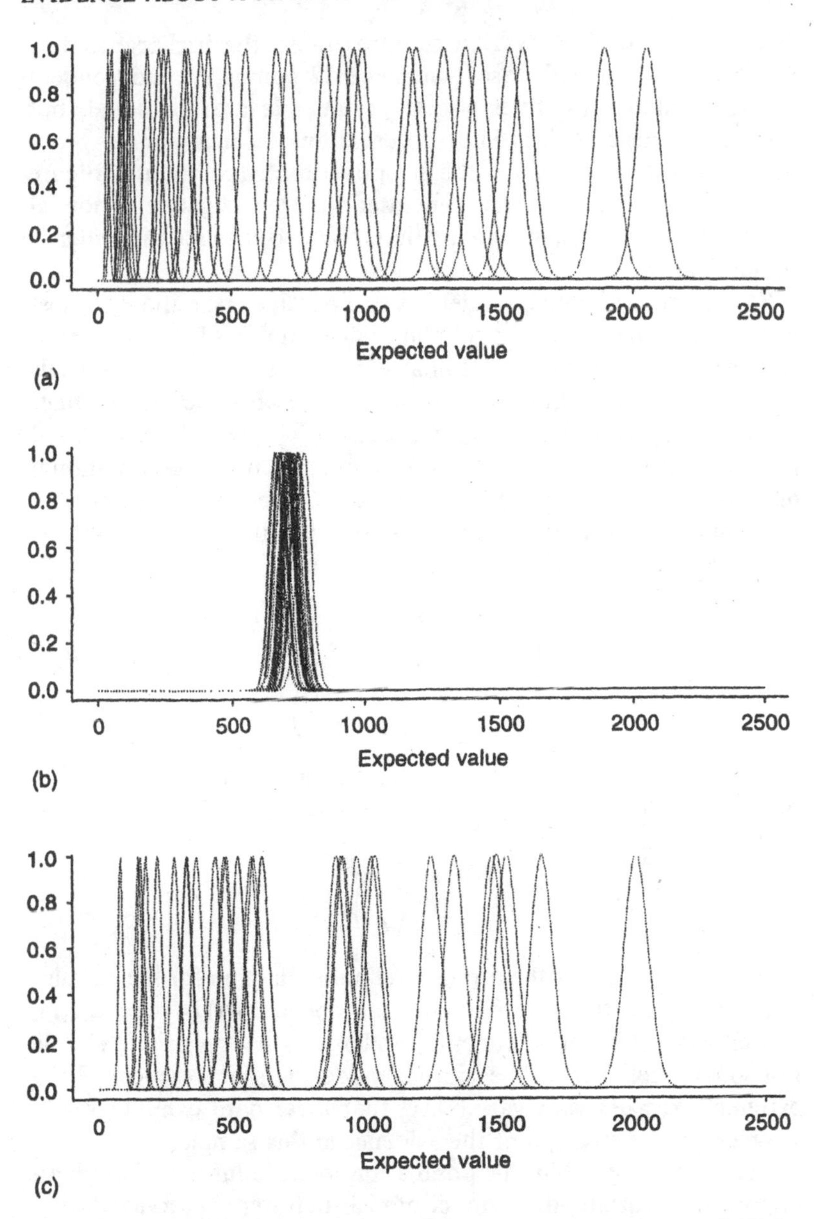

Figure 6.5 *Likelihoods for Poisson parameters: (a) hospital discharges; (b) Poisson data; (c) negative binomial data.*

sample mean, while the variance approximates the discharge sample variance. (This model arises when the 32 counts are independent Poisson random variables whose parameters are not equal, but instead are generated as i.i.d. observations from a gamma probability distribution.) Since the hospital discharge counts (Figure 6.5(a)) look like the data generated by the negative binomial model (Figure 6.5(c)), we will use that model to examine the evidence in our sample.

The negative binomial model has two parameters, r and θ; it most commonly appears as the probability distribution of the number of failures before the rth success in a sequence of i.i.d. Bernoulli trials with success probability θ. Under this model, when the total number of discharges from the sample hospitals is $\sum y_i = t_s$, the likelihood function for the population total, t, is obtained from the conditional distribution of the vector of sample counts, given t. The parameter θ conveniently drops out, so the likelihood is a function of t and the nuisance parameter r:

$$L(t,r) \propto \frac{\binom{t - t_s + (N-n)r - 1}{(N-n)r - 1} \prod_{i=1}^{n} \binom{y_i + r - 1}{r - 1}}{\binom{t + Nr - 1}{Nr - 1}}.$$

By contrast, the likelihood under the simple Poisson model is proportional to the hypergeometric probability:

$$L(t) \propto \frac{1}{\binom{t}{t_s}}.$$

Figure 6.6(a) shows the profile likelihood function for the population total under the negative binomial model, $L_P(t) = \max_r L(t,r)$, as well as the likelihood function under the (inappropriate) simple Poisson model. Using the simple model in the presence of the extreme 'extra-Poisson variability' that these data exhibit greatly exaggerates the strength of the evidence in this sample.

For every hospital in this population we have information on an important covariate, the number of beds that were in service during the time of the study. We can use this information to construct an alternative model for the sample, representing the observation from hospital i as a normal random variable whose expected value and variance are both proportional to the number of beds, x_i. (Shortcomings of this model for representing the relationship

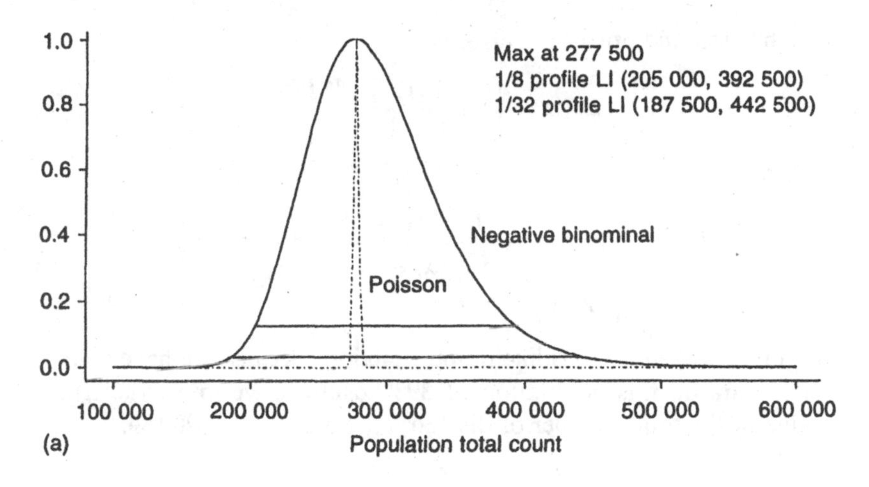

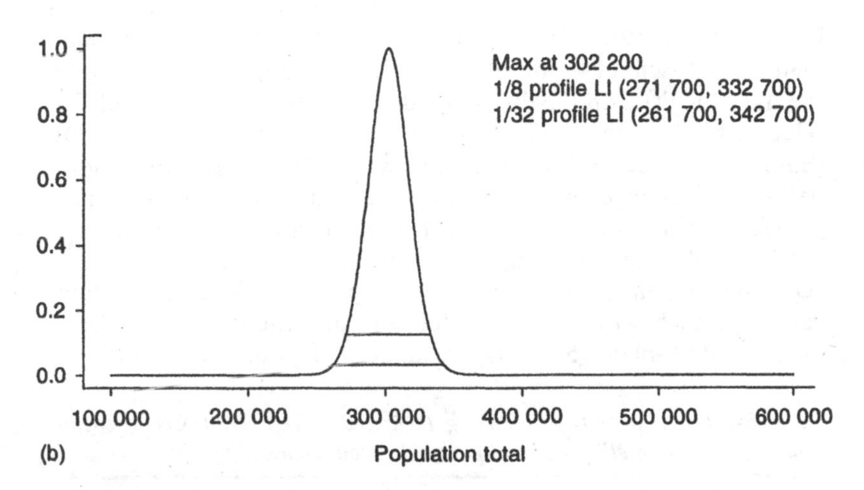

Figure 6.6 *Likelihoods for total discharges from 393 hospitals: (a) negative binomial and Poisson models; (b) normal regression model.*

between the number of discharges and the number of beds in this particular population have been examined in some detail from the viewpoint of standard estimation and confidence interval methodology: see Royall and Cumberland, 1981a; 1981b; 1985). If we denote by x_i the number of beds in hospital i, by $\bar{x}_s$ the average number of beds in the 32 sample hospitals, and by $\bar{x}$ the average number in all 393 hospitals, then the profile likelihood function under this model – modified by using $n - 1$ instead of n in the exponent, as suggested by

Kalbfleisch and Sprott (1970) – is

$$L_P(t) \propto \left[1 + \frac{T(t)^2}{n-1}\right]^{-(n-1)/2},$$

where

$$T(t) = \frac{t - (\bar{y}_s/\bar{x}_s)N\bar{x}}{\sqrt{\frac{N^2}{n}\left(1 - \frac{n}{N}\right)\frac{\bar{x}\bar{x}_r}{\bar{x}_s}\sum_s \frac{(y_i - (\bar{y}_s/\bar{x}_s)x_i)^2}{(n-1)x_i}}}.$$

Figure 6.6(b) shows this likelihood function. Because we have complete data on this population of 393 hospitals, we know the true value of the total number of discharges, which is $t = 320\,159$.

6.6 Determinants of plans to attend college

Cox and Snell (1981) and Snell (1987) used data on the educational plans of high-school senior boys (Sewell and Shah, 1968) to illustrate the use of logistic and loglinear models. The study investigated the relationship between socioeconomic status (SES), intelligence (IQ) (both at four levels: low, medium low, medium high, and high), parental encouragement (low, high), and the response variable, plans for attending college (yes, no). The data are given in Table 6.4.

We model the responses within each of the 32 SES × IQ × encouragement subgroups as independent binomial random variables, each with its own value for the probability of college plans, and display 16 of the likelihood functions in Figure 6.7

Table 6.4 *Fraction of boys planning to attend college, by levels of socio-economic status, intelligence, and parental encouragement*

IQ	Parental encouragement	SES L	SES ML	SES MH	SES H
L	Low	4/353	2/234	8/174	4/52
	High	13/77	27/111	47/138	39/96
ML	Low	9/216	7/208	6/126	5/95
	High	33/105	64/159	74/184	123/213
MH	Low	12/138	12/127	17/109	9/50
	High	38/92	93/185	148/248	224/289
H	Low	10/77	17/96	6/48	8/25
	High	49/92	119/178	198/271	414/468

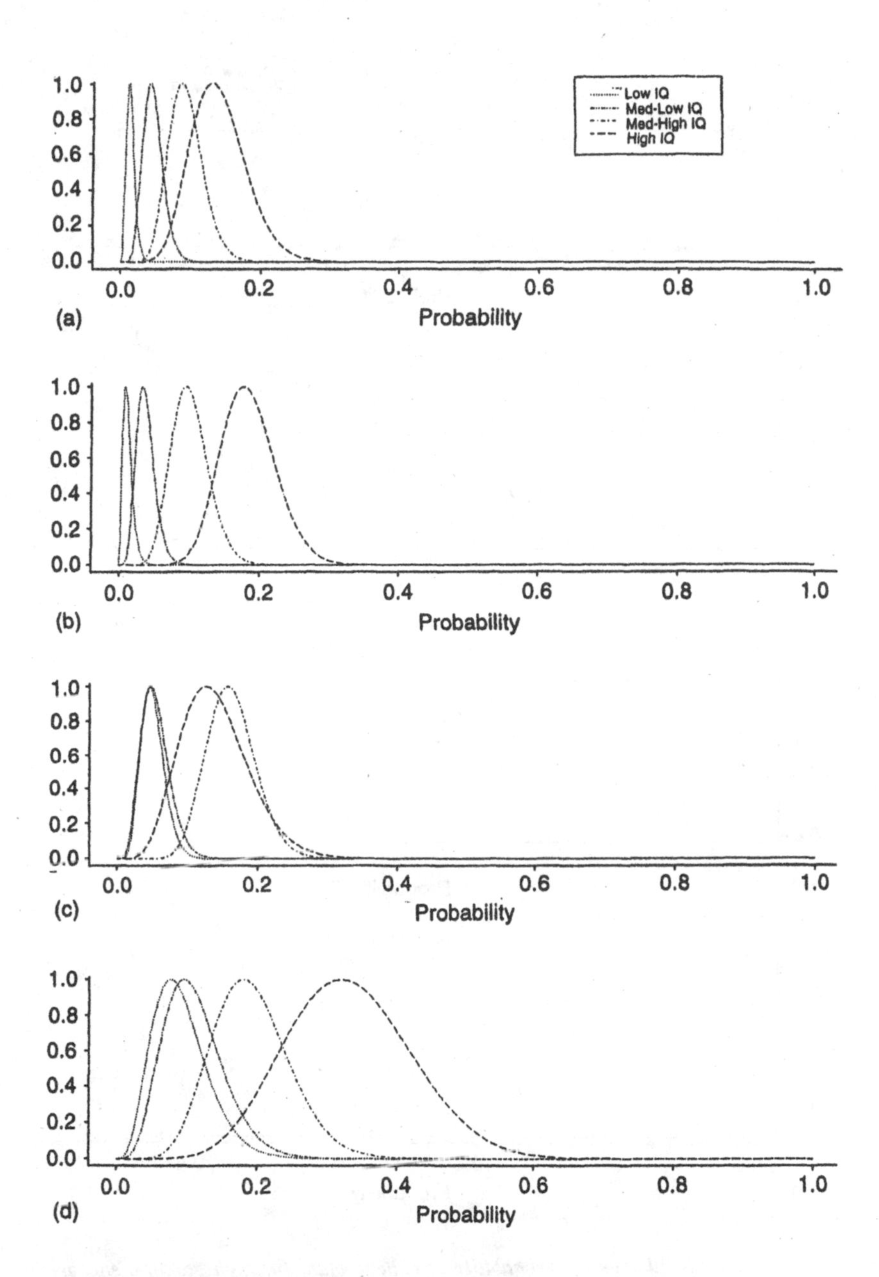

Figure 6.7 *Likelihoods for probability of college plans for low parental encouragement: (a) SES low; (b) SES medium low; (c) SES medium high; (d) SES high.*

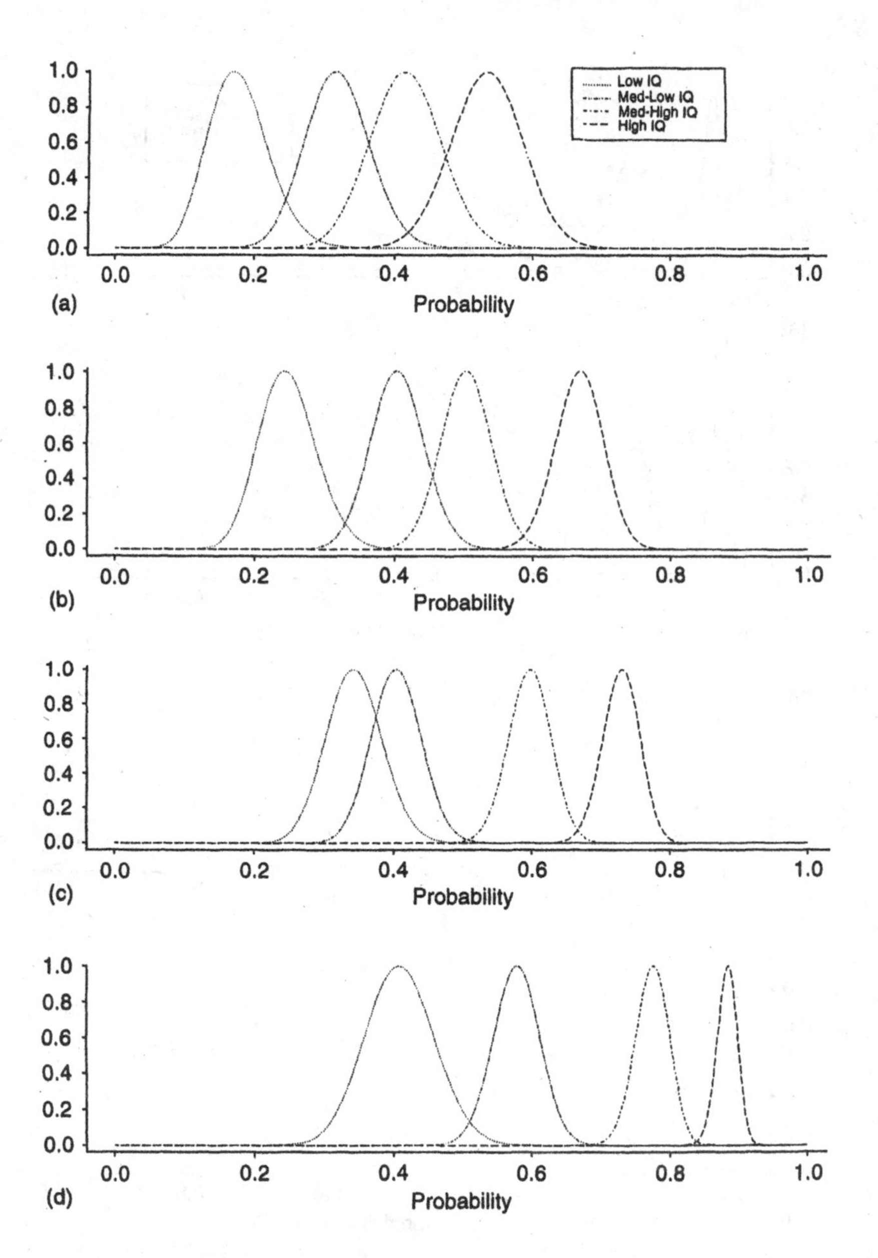

Figure 6.8 *Likelihoods for probability of college plans for high parental encouragement: (a) SES low; (b) SES medium low; (c) SES medium high; (d) SES high.*

(encouragement low) and 16 in Figure 6.8 (encouragement high). These two graphs represent all of the evidence in this data set under the present model (32 binomials). Examination of the likelihood functions for parameters in the logistic model would further help to quantify the evidence, but these two figures show clearly the strong and consistent effects of SES and IQ, and the striking additional effect of parental encouragement, on the probability that a boy plans to attend college.

6.7 Evidence about probabilities in a $2 \times 2 \times 2 \times 2$ table

To illustrate and compare various methods for analyzing multiple 2×2 tables, Gart (1971) and Gart and Nam (1988) used data from a controlled study of the possible carcinogenic effect of a fungicide on mice. Two strains of mice were used (X and Y), so that there were four sex-by-strain subgroups. The data are shown in Table 6.5.

First, we examine the evidence under a general model that treats the number of tumor mice in each of the eight treatment-by-sex-by-strain groups as independent binomial random variables, each with its own value for the parameter (probability of tumor). Figure 6.9 separates the likelihoods for the four treatment groups from the four control groups. These two panels represent all of the evidence in this data set under this model. It is clear at a glance that these data support substantially higher values for the tumor probabilities in the treatment groups.

To quantify the effect of treatment within each of the four sex-by-strain groups we first use the odds ratio, that is, the ratio of the odds on tumors for treated mice to the odds for controls. To see how the effect might vary over these four groups, we look at the likelihood

Table 6.5 *Frequency of tumors in four subgroups of mice*

Subgroups		Tumor	No tumor
X males	Treated	4	12
	Control	5	74
X females	Treated	2	14
	Control	3	84
Y males	Treated	4	14
	Control	10	80
Y females	Treated	1	14
	Control	3	79

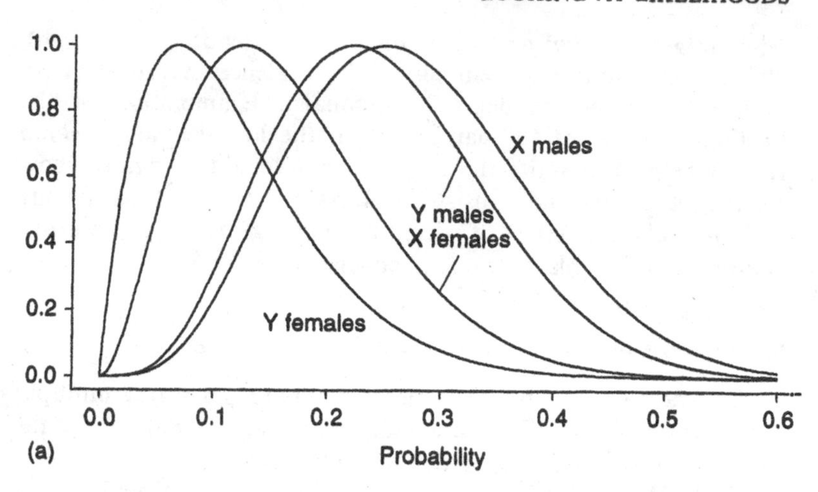

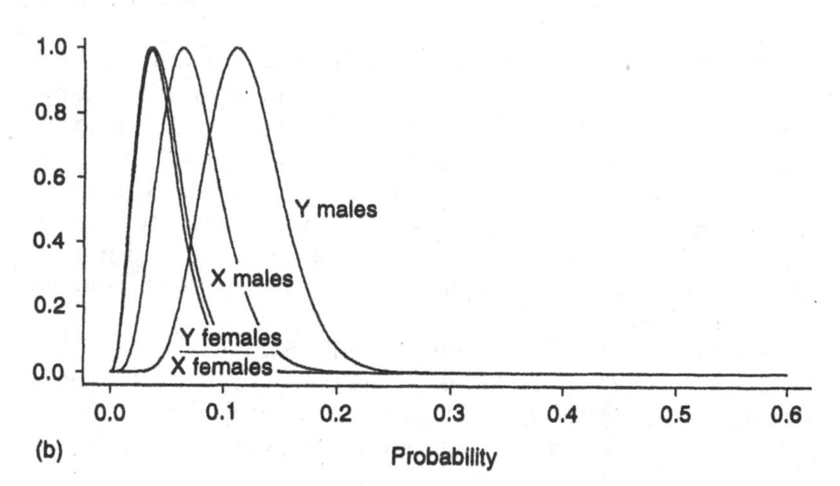

Figure 6.9 *Likelihoods for tumor probability: (a) treated mice; (b) control mice.*

function for the odds ratio in each group (the light lines in Figure 6.10), as well as at the likelihood function under the simple model that says one common odds ratio applies to all four (the heavy line). Odds ratio likelihoods are obtained from the conditional distributions, given total numbers of tumor mice in each subgroup. The likelihood under the simple model is proportional to the product of the four subgroup likelihoods.

The four subgroup curves are quite compatible, and the simple model curve, whose numerical description appears in the upper

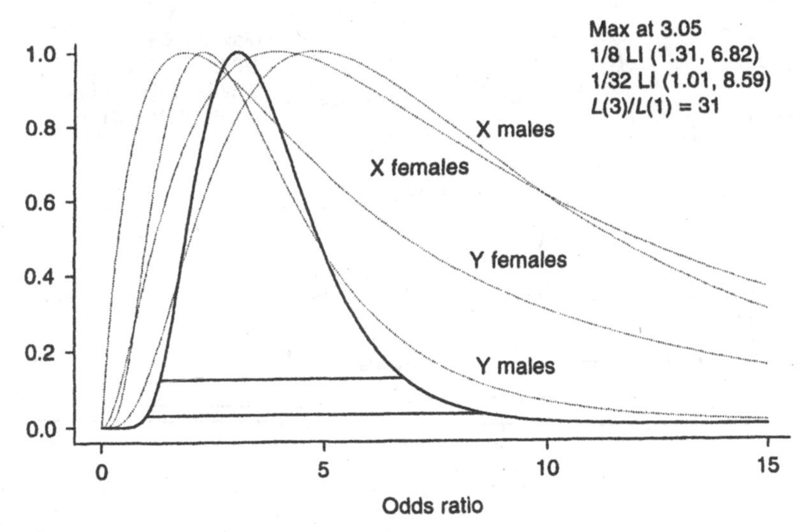

Figure 6.10 *Likelihoods for ratio of odds on tumor: treated versus control mice.*

right-hand corner, shows that these observations represent strong evidence supporting odds ratios of 3 or so versus the value, 1, that corresponds to no treatment effect ($L(3)/L(1) = 31$).

Another measure of the effect of treatment is the probability ratio – that is, the ratio of the probability of tumors in treated mice to the probability in controls. Again we look at the evidence in each of the four sex-by-strain groups (Figure 6.11, light lines) as well as under the simple model that says one probability ratio (heavy line) applies to all four (cf. Gart and Nam, 1988). (These probability ratio likelihoods are synthetic conditional likelihoods, described in detail in section 7.7. The likelihood under the simple model is again proportional to the product of the four subgroup likelihoods.) The overall picture is similar to that for the odds ratio (Figure 6.10), and we will use the odds ratio as a measure of treatment effect to further examine this data set.

Next we look at the evidence about the effects of sex and strain, and at how the treatment effect might differ according to these characteristics. First we look at the ratio of tumor odds in males to the odds in females in each of the four treatment-by-strain subgroups (Figure 6.12). Again we see that the evidence from the four subgroups is quite compatible with a simple model stating that they have a common odds ratio. There is fairly strong evidence that the males' odds in favor of tumor is two to three times the females' odds ($L(2.5)/L(1) > 20$). (Figure 6.10 showed that there

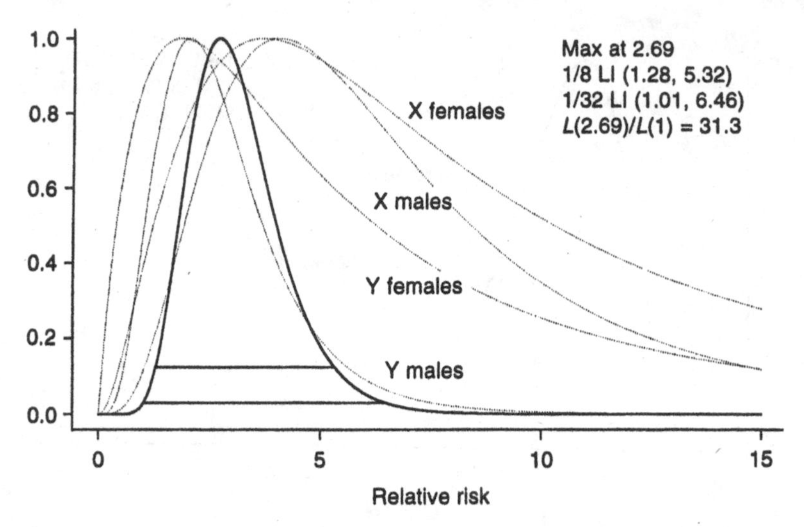

Figure 6.11 *Likelihoods for ratio of probabilities of tumor: treated versus control mice.*

is essentially no evidence that the treatment affects the male and female tumor odds differently.)

Figure 6.13 shows the evidence about the strain effect. It consists of the likelihoods for the ratios of tumor odds in strain Y versus strain X mice in each of the four treatment-by-sex subgroups, plus the likelihood under a common-odds-ratio model. There is no

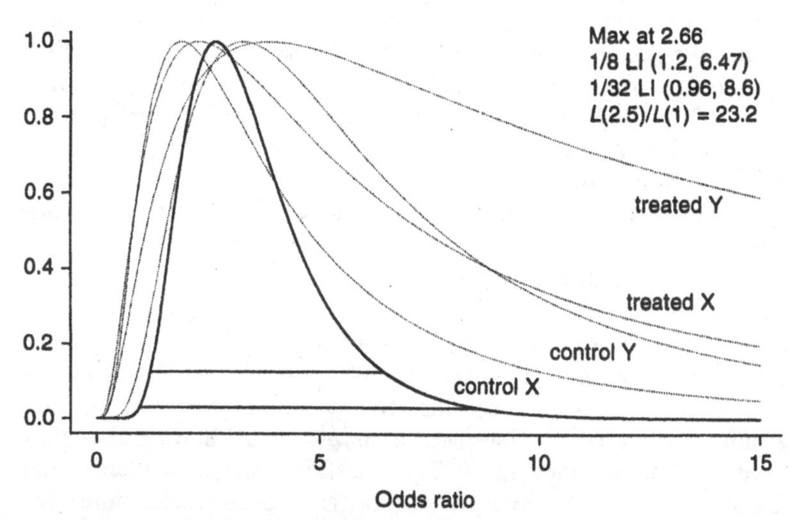

Figure 6.12 *Likelihoods for ratio of odds on tumor: males versus females.*

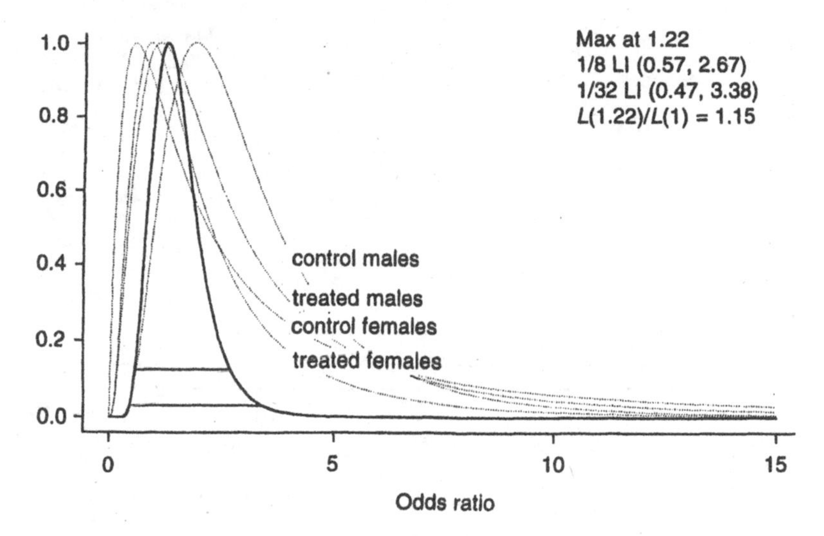

Figure 6.13 *Likelihoods for ratio of odds on tumor: strain Y versus strain X.*

value of the common odds ratio that is much better supported than the no-strain-effect value, 1. (The maximum is at an odds ratio of 1.22, and the support for that value versus 1 is only $L(1.22)/L(1) = 1.15$.)

Nevertheless, it appeared in Figure 6.10 that most of the evidence for a strong treatment effect comes from the strain X mice. This is made clearer in Figure 6.14, whose two panels compare the evidence about the treatment-versus-control odds ratio in strain X mice to the evidence in strain Y mice. In these panels we have displayed the numerical values of the likelihood ratios in favor of 3, the approximate overall maximum (Figure 6.10) versus 1 in order to see the respective contributions from the two strains. In strain Y odds ratios of 2–3 are better supported than 1, but the evidence is weak. (The maximum is at 2.16, and $L(2.16)/L(1)$ is only 2.3.)

6.8 Evidence from a community intervention study of hypertension

Swan (1993) demonstrated how GLIM software can be used to construct a complete analysis of variance from summary data, using the results of a study in which one community was exposed to a campaign aimed at reducing hypertension, while another community was simply observed as a control. Data on systolic blood pressure for samples from both communities at the end of the study are given in Table 6.6.

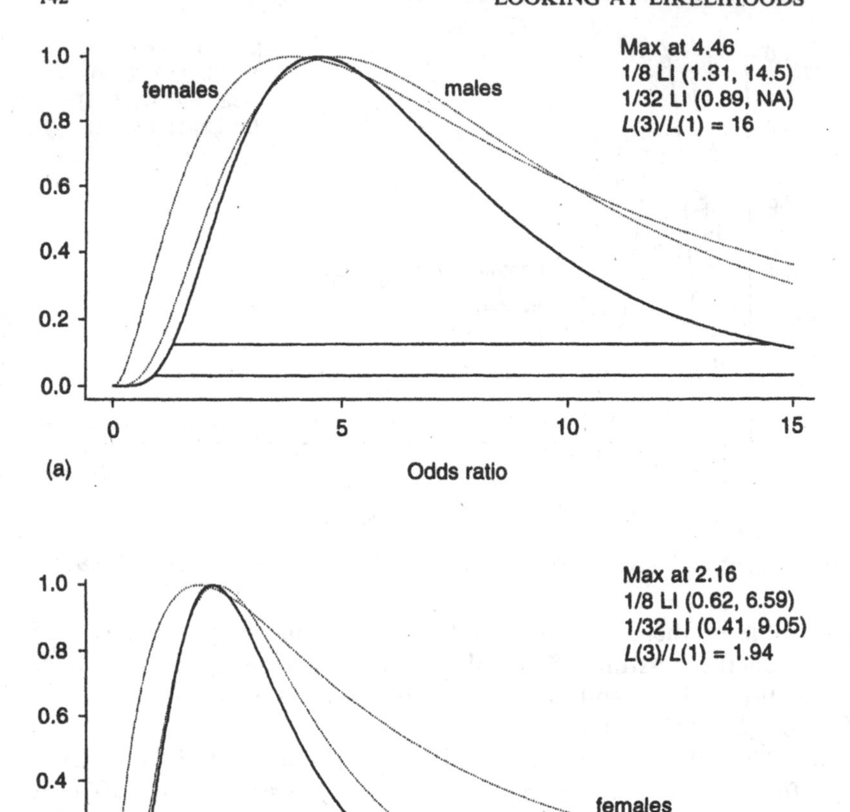

Figure 6.14 *Likelihoods for ratio of odds on tumor: treated versus control mice: (a) strain X; (b) strain Y.*

Under a normal distribution model, profile likelihoods for the mean blood pressure of males and of females in the two communities are shown in Figure 6.15. (These are profile likelihoods modified by replacing the sample size n by $n - 2$ in the exponent – see Cox and Reid (1987) for one rationale for this modification.) There is very strong evidence supporting the hypothesis that males in the intervention community had mean pressure about 5% lower than males in the control community (versus the hypothesis of no difference),

Table 6.6 *Systolic blood pressure data*

		Control community	Intervention community
Males	n	200	223
	mean	158.5	150.1
	s.d.	24.2	21
Females	n	191	283
	mean	167.4	154.7
	s.d.	27.5	21.5

and even stronger evidence for an even greater difference in females. As the profile likelihoods in Figure 6.16 show, a male intervention/control ratio of 0.95 is better supported than a ratio of 1 by a factor of well over 1000. A female ratio of 0.95 is better supported than 1 by almost 400 000, and there is some evidence that the female ratio is even smaller – $L(0.924)/L(0.95) = 6.7$.

A measure of the difference between the intervention's effects on males and females could be represented by the difference between the females' intervention/control ratio and the males' (or by the ratio of these two ratios). A profile likelihood function for one of these quantities would synthesize the evidence about the two individual ratios (represented in the two curves in Figure 6.16) and provide a simple direct expression of the evidence about how the treatment's effects on men and women differ. Software for drawing these profile likelihoods should soon be available.

There is very strong evidence here that females' mean blood pressure is 5–7% higher than males' in the control community. Figure 6.17 shows that $L(1.05)/L(1)$ is over 300. There is somewhat weaker evidence for a smaller ratio in the intervention community, where the likelihood is maximized at 1.03, and $L(1.03)/L(1) = 18.4$. This is consistent with the previous observation that these data represent some evidence that the treatment/control ratio is smaller in women than in men, that is, evidence that if the intervention is the cause of the blood-pressure differences between these two communities, then it is more effective in women than in men.

6.9 Effects of sugars on growth of pea sections: analysis of variance

An experiment in plant physiology studied how the growth of pea sections in tissue culture might be affected by the addition of various

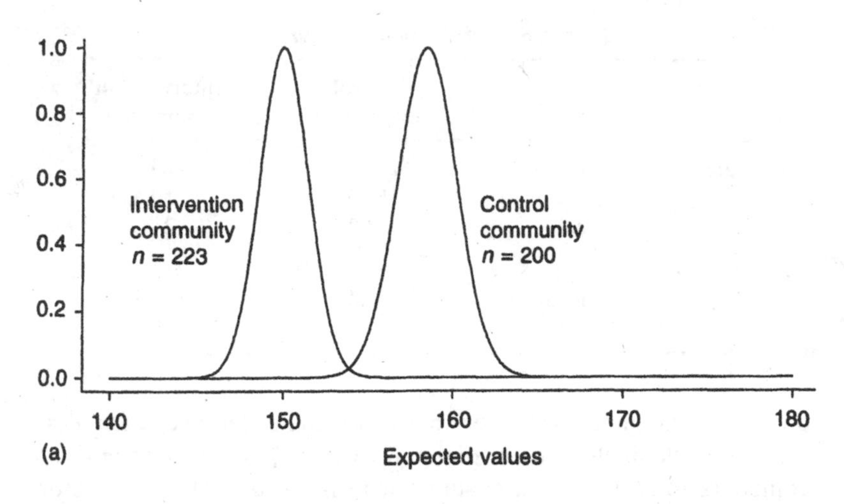

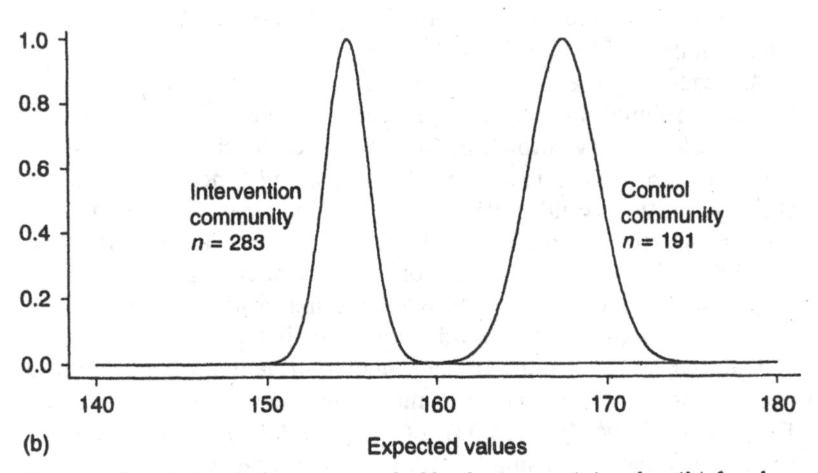

Figure 6.15 *Likelihoods for mean systolic blood pressure: (a) males; (b) females.*

sugars. There were four experimental groups, representing three different sugars and one sugar mixture, plus a control group. Ten replications were made for each of the five groups, with results shown in Table 6.7. Sokal and Rohlf (1969) used these data to illustrate Model I analysis of variance and the Student–Newman–Keuls multiple comparisons procedure.

Figure 6.18 shows that this experiment produced very strong evidence that sucrose reduces growth sharply, and that glucose, fructose, and their mixture reduce it even more. (Again, these profile

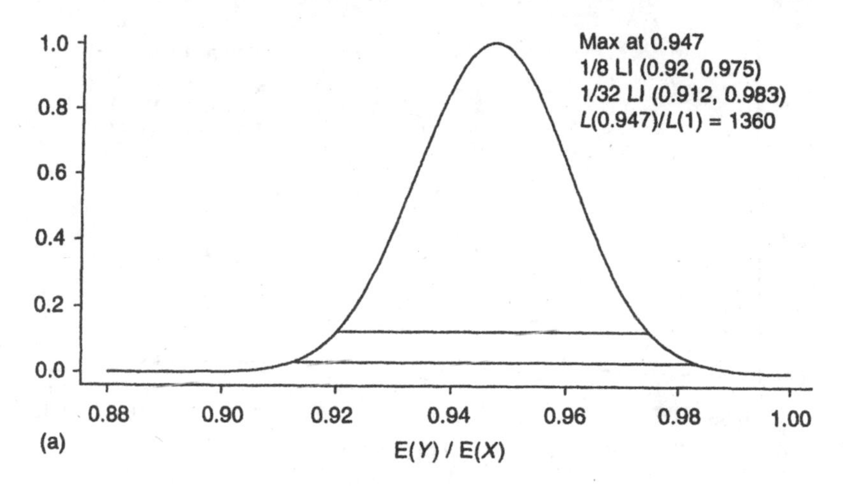

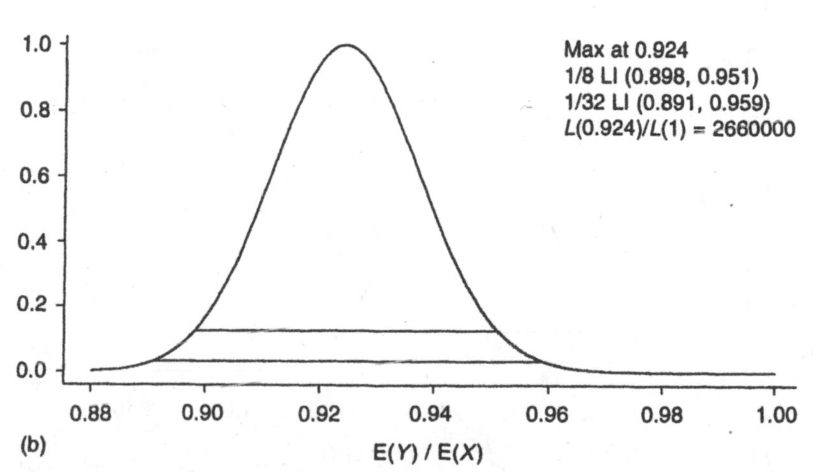

Figure 6.16 *Likelihoods for ratio of mean systolic blood pressure: (a) intervention/control males; (b) intervention/control females.*

likelihoods are modified by replacing the sample size n by $n - 2$ in the exponent.) There is little evidence for differences among these last three treatments. This figure exhibits the primary evidence about treatment effects that the experiment was designed to produce.

A more detailed analysis of the variability among these observations appears in Figure 6.19, which shows estimated likelihood functions for each of the 50 observations. These are simply single-observation likelihoods for a normal mean, with the variance

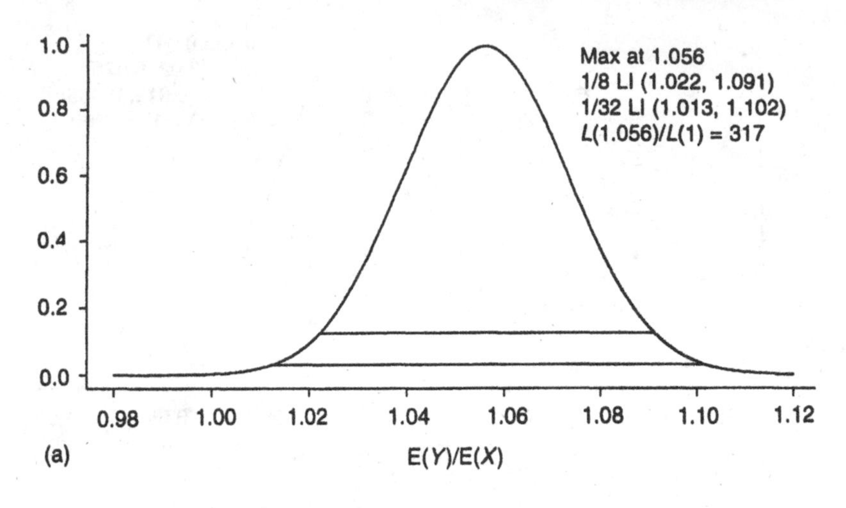

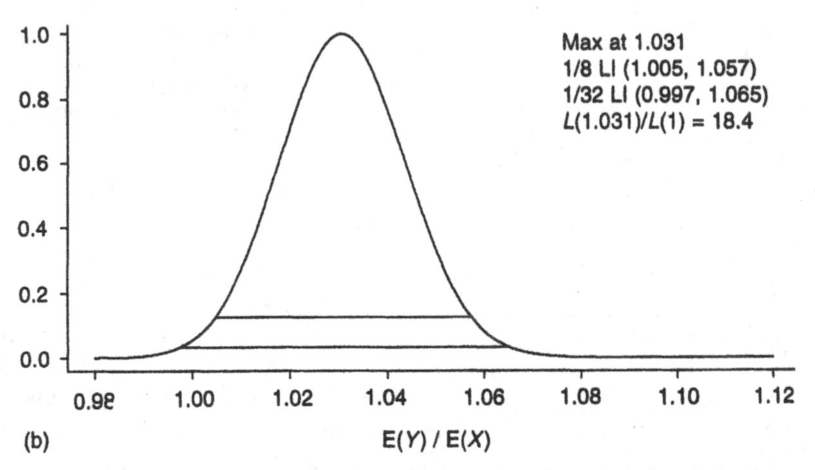

Figure 6.17 *Likelihoods for ratio of mean systolic blood pressure, females/males: (a) control community; (b) intervention community.*

parameter estimated for observations in each treatment group by the sample variance within that group.

Finally, for an explicit representation of the evidence about the effect of sucrose, we look at Figure 6.20, which shows that values for the ratio of the control mean to the sucrose mean in the range 1.06–1.13 are much better supported than the null value, 1 ($L(1.06)/L(1) > 5230/8 = 665$, stronger evidence than that in nine consecutive white balls in the urn scheme of Chapter 1).

Table 6.7 *Length of pea sections grown in tissue culture*

Treatments				
Control	2% glucose	2% fructose	2% sucrose	1% glucose + 1% fructose
75	57	58	62	58
67	58	61	66	59
70	60	56	65	58
75	59	58	63	61
65	62	57	64	57
71	60	56	62	56
67	60	61	65	58
67	57	60	65	57
76	59	57	62	57
68	61	58	67	59

We note that the likelihood analysis avoids the complications presented by the larger variance in the control group and the issue of whether a particular comparison between means was specified a priori, as well as how one might interpret the finding of the analysis of variance that 'mixed sugars affect the sections differently from pure sugars' (Sokal and Rohlf, 1969, p. 229).

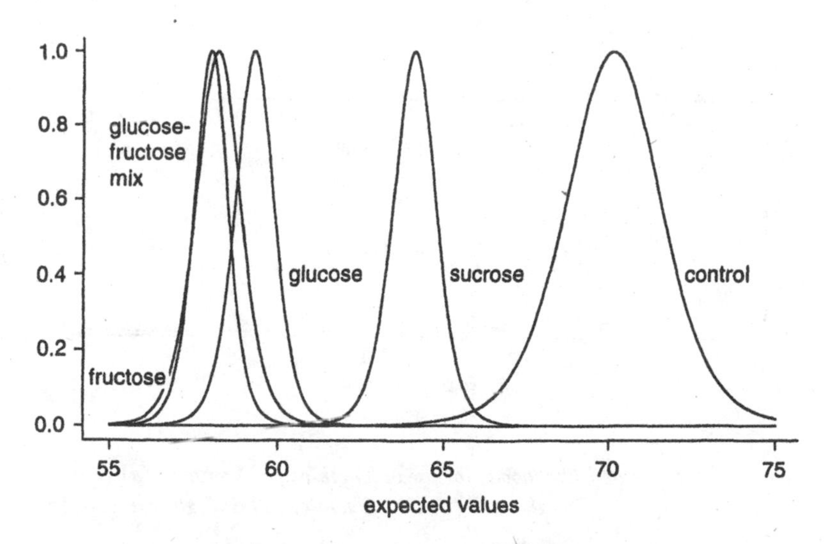

Figure 6.18 *Likelihoods for mean length of pea sections.*

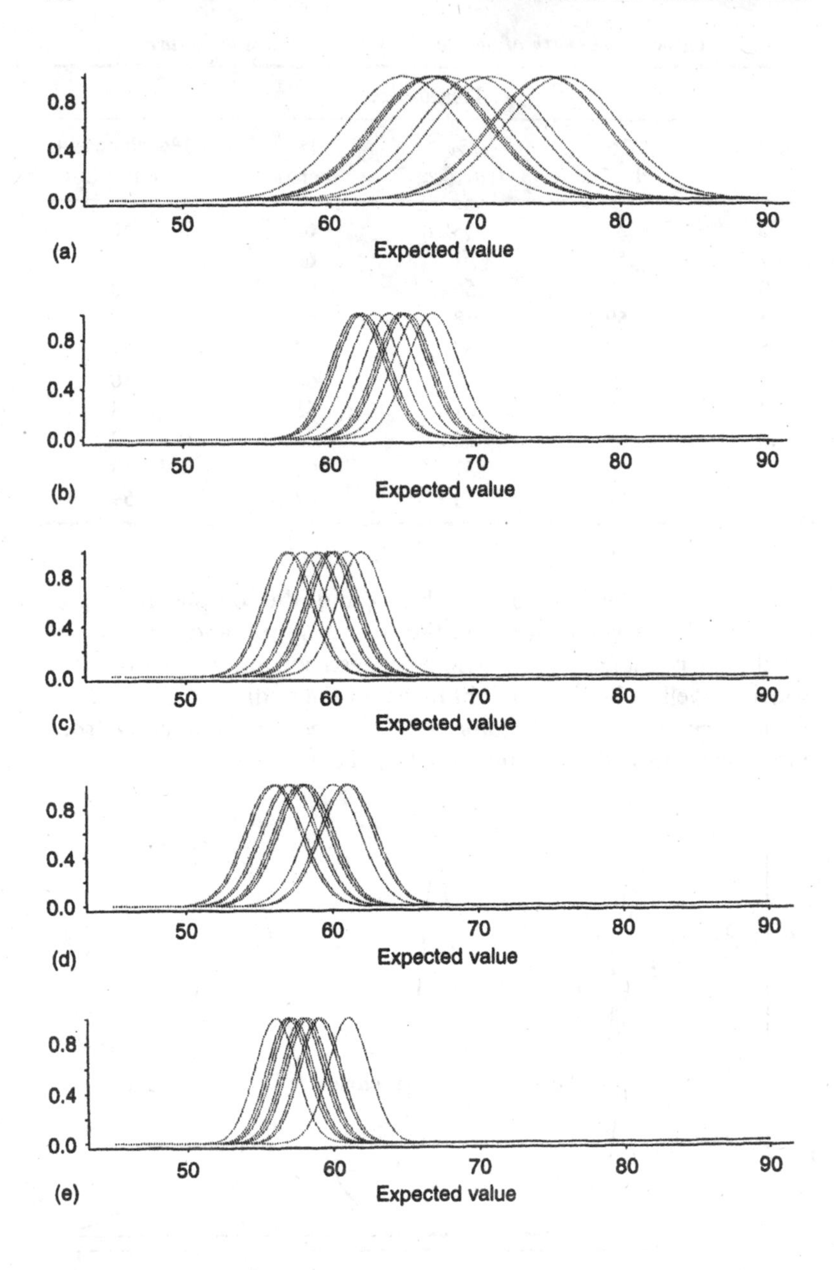

Figure 6.19 *Estimated likelihoods for mean length of pea sections: (a) control; (b) 2% sucrose; (c) 2% glucose; (d) 2% fructose; (e) 1% glucose plus 1% fructose.*

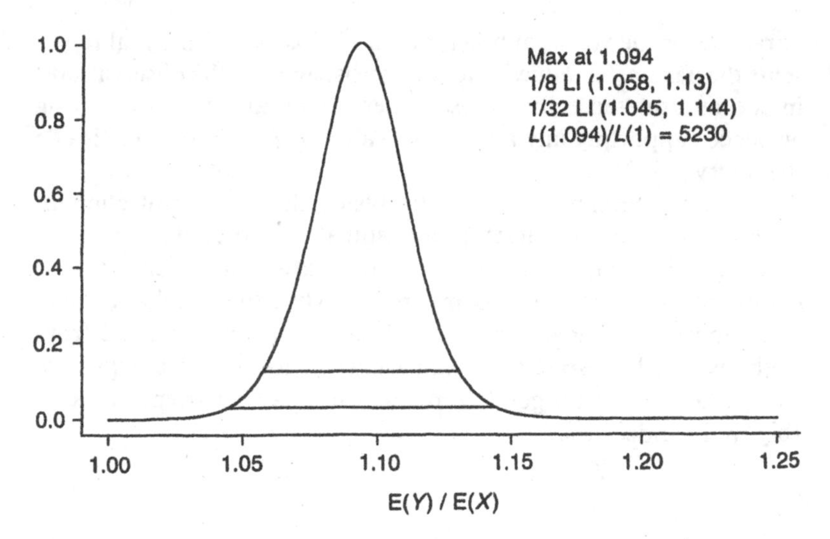

Figure 6.20 *Profile likelihood for ratio of normal means: control versus sucrose.*

6.10 Summary

Note that in this chapter, which is concerned with the interpretation of data as evidence, the probabilities of weak and of misleading evidence played no role. These probabilities do not affect the interpretation of likelihood functions. Calculating the probabilities of finding weak or misleading evidence, for various combinations of parameter values, is important for *planning* a study. These calculations give a good idea of what to expect if a particular sample size or sampling strategy is adopted. For various reasons, such as budgeting time and money, it may be wise to adopt a specific sampling plan, but a plan is nothing more. It is not a contract whose alteration carries a statistical penalty. Planning to make a certain number of observations does not preclude stopping earlier, if the evidence turns out to be stronger than expected, or continuing beyond the planned sample size, if stronger evidence is required. When the study is done, the evidential meaning of the observed data is quite independent of such contingencies, and is expressed *in toto* by likelihood functions.

The way to see what the data say about a parameter of interest is to look at the entire likelihood function for that parameter. Examining specific parts of the likelihood function can serve purposes for which conventional frequentist methods are often used. In particular, likelihood ratios measure the strength of evidence for one parameter value versus another, and a 1/8 likelihood interval represents the parameter values that are 'consistent with the observations' in the natural sense that those observations are not very strong evidence supporting any alternative value *vis-à-vis* any value inside the interval.

In many common statistical problems the usual collection of methods – significance tests, point estimates, and confidence intervals – gives a general impression of the statistical evidence that is reasonably accurate, when compared to what the likelihood function shows. Of course, this is just what should be expected from methods which, despite their well-known theoretical inadequacies, have, when used with good common sense, served science reasonably well for decades.

CHAPTER 7

Nuisance parameters

7.1 Introduction

We can plot the likelihood function for a model whose distributions are indexed by a one-dimensional parameter θ and literally see what the data say. We might take note of special features, such as where the function is maximized, the limits of the 1/8 likelihood interval, etc., but the best way to see what the data say, in a general way, is simply to look at the entire likelihood function.

When the parameter is multi-dimensional, it is not as easy to appreciate the likelihood function. The meaning is the same – the distribution identified by the parameter value θ_1 is better supported than the one identified by θ_2 by the factor $L(\theta_1)/L(\theta_2)$, regardless of the dimension of θ. But as the dimension increases, the function becomes difficult to visualize and understand. This is especially troublesome when we are primarily interested in one aspect of the model, represented by one component, or a one-dimensional function, of the parameter vector, such as one coefficient in a multiple regression model or the odds ratio in a two-binomial model. And even when we are interested in an entire multi-dimensional parameter vector, we have a great advantage if we can represent, interpret, and communicate the evidence about its components one, or a few, at a time. Thus, even though we might well have been interested in all ten of the parameters in the model that we used for the pea data in section 6.9 (five normal distributions), we sought first to examine the evidence about each of the five means without simultaneously worrying about the variances (Figure 6.18), and then to look at only the evidence about a ratio of two means, without worrying about their individual values or the two variances (Figure 6.20).

For discussing this issue it is convenient to change our notation and to use distinct symbols (θ, γ) for the two components of the parameter vector, where θ is now the part that we are interested in at the moment, and γ is the rest. Our problem is that we want to represent, interpret, and report the evidence about θ alone, but the likelihood function $L(\theta, \gamma)$ depends on both θ and the nuisance parameter γ. The likelihood ratio $L(\theta_1, \gamma)/L(\theta_2, \gamma)$ measures the relative support for the two probability distributions identified by (θ_1, γ) and (θ_2, γ) or, equivalently, the support for the parameter value (θ_1, γ) versus the value (θ_2, γ). For a fixed value of γ this ratio measures the support for θ_1 versus θ_2, but it depends on the value of γ. We want a likelihood function for θ alone, but in general we cannot remove the nuisance parameter γ. This is because it is the vector (θ, γ), not θ alone, that determines a probability distribution for the observed random variable, and what likelihood ratios measure is support for one *distribution* versus another. A value of θ does not determine a distribution, but a *family* of distributions (defined as γ varies), and we saw in section 1.9 that we cannot, in general, compare and measure the support for families of distributions.

There is no general solution to this problem of nuisance parameters. There is no general way to eliminate them that is theoretically unassailable. But, as the examples in Chapter 6 have shown, there are quite satisfactory *ad hoc* methods for many important applications. In this chapter we consider some of the strategies that have been proposed for eliminating nuisance parameters, and see the solutions that they provide for some fundamental probability models.

7.2 Orthogonal parameters

Suppose we have observed independent random variables X and Y, where the distribution of X depends only on θ and that of Y depends only on γ, with θ and γ unrelated. In this case the entire likelihood function consists of two factors, $L(\theta, \gamma) \propto f(x; \theta)g(y; \gamma)$, and the first of these, $f(x; \theta)$, is the likelihood function for θ. The relative support for any two values of θ is $L(\theta_1, \gamma)/L(\theta_2, \gamma) = f(x; \theta_1)/f(x; \theta_2)$ regardless of γ. More generally, whenever the likelihood function has this same product form, $L(\theta, \gamma) = L_1(\theta)L_2(\gamma)$, the parameters θ and γ are said to be **orthogonal** (Anscombe, 1964), and the factor $L_1(\theta)$ is the likelihood function for θ.

Sometimes the parameters are not orthogonal, but orthogonality can be achieved by reparameterizing the model – that is, we can orthogonalize. For example, if X_1 and X_2 are independent Poisson variables with parameters λ_1 and λ_2 and we are interested in the ratio $\theta = \lambda_1/\lambda_2$, we can reparameterize in terms of θ and the nuisance parameter $\gamma = \lambda_1 + \lambda_2$, so that $\lambda_1 = \theta\gamma/(1 + \theta)$,

$\lambda_2 = \gamma/(1+\theta)$, and the likelihood function is

$$L(\theta,\gamma) \propto \theta^{x_1}/(1+\theta)^{x_1+x_2}\gamma^{x_1+x_2}\,e^{-\gamma}. \tag{7.1}$$

The evidence about the ratio $\theta = \lambda_1/\lambda_2$ is given by the factor $\theta^{x_1}/(1+\theta)^{x_1+x_2}$, and the evidence about the sum $\gamma = \lambda_1 + \lambda_2$ is given by $\gamma^{x_1+x_2}\,e^{-\gamma}$.

An important case where orthogonalization is possible is the multivariate normal (MVN) distribution model with variable mean vector μ and fixed covariance matrix Σ. The likelihood function for any component of the mean vector can be obtained by reparameterizing the model in terms of a vector parameter (θ,γ) having the same dimension as μ, where θ is the component of interest and is orthogonal to γ. This likelihood function for θ turns out to be simply the one based on the marginal normal distribution of the corresponding component of the observed random vector (Exercise 7.3). Thus if X is a p-dimensional random vector with a $MVN(\mu,\Sigma)$ distribution, the likelihood function for the second component, μ_2, is proportional to $\exp\{-(x_2-\mu_2)^2/2\sigma_2^2\}$, where σ_2^2 is the variance of X_2, that is, the second diagonal element of Σ.

The same result holds for any linear function of the elements of the mean vector μ – the likelihood function for $\Sigma_1^p \ell_j\mu_j = \ell'\mu$ alone can be obtained by reparameterizing in terms of $\ell'\mu$ and an orthogonal nuisance parameter. This likelihood function turns out to be simply the one generated by the $N(\ell'\mu, \ell'\Sigma\ell)$ model for the marginal distribution of $\ell'X$ – that is, it is proportional to $\exp\{-(\ell'x-\ell'\mu)^2/2\ell'\Sigma\ell\}$.

For a specific coefficient in a normal linear regression model we can rewrite this last result in an alternative familiar form. Suppose Y is MVN with $EY = X\beta + Z\gamma$, where X is an $n \times 1$ vector of covariates associated with the real-valued coefficient of interest, β, and Z is a matrix of regressors whose coefficients, the elements of the vector γ, are nuisance parameters. When the covariance matrix Σ equals $I\sigma^2$ the likelihood function for β is proportional to the $N(\hat\beta, \text{var}(\hat\beta))$ density, $\exp\{-(\hat\beta-\beta)^2/2\,\text{var}(\hat\beta)\}$, where if D denotes the matrix $I - Z(Z'Z)^{-1}Z'$, then $\hat\beta = (X'DY)/X'DX$ and $\text{var}(\hat\beta) = \sigma^2/X'DX$. (In the notation of the preceding paragraph, $\mu = X\beta + Z\gamma$ and $\ell' = X'D$, so that $\ell'\mu = \beta$ is the parameter of interest.) This same result applies for a general covariance matrix Σ if X, Y, and Z are replaced by $\Sigma^{-1/2}X$, $\Sigma^{-1/2}Y$, and $\Sigma^{-1/2}Z$, respectively, and D becomes $\Sigma^{-1} - \Sigma^{-1/2}Z(Z'\Sigma^{-1}Z)^{-1}Z'\Sigma^{-1/2}$.

Other important cases where orthogonalization is possible are described in Exercises 7.1 and 7.5. These pertain to sums and ratios of expected values of multiple independent Poisson random variables and to sums of cell probabilities in the multinomial distribution.

7.3 Marginal likelihoods

When orthogonal reparameterization is not possible, we still can find useful representations of the evidence for a parameter of interest without nuisance parameters. For example, consider the $N(\mu, \sigma^2)$ model when we are interested in the evidence about the variance σ^2 in a sample $(x_1, \ldots, x_n)$, so that μ is a nuisance parameter. The parameters μ and σ^2 are not orthogonal, since the likelihood function is $L(\mu, \sigma^2) \propto (\sigma^2)^{-n/2} \exp\{-\Sigma(x-\mu)^2/2\sigma^2\}$ and an exact orthogonal reparameterization is not possible. But the statistic $S^2 = \Sigma(X - \bar{X})^2/(n-1)$ has a marginal probability distribution that depends on σ^2, and is free of the nuisance parameter μ. Therefore the observation $S^2 = s^2$ provides evidence about σ^2 that is independent of μ, and that evidence is represented by the marginal likelihood function,

$$L_M(\sigma^2) \propto f_S(s^2; \sigma^2) \propto (\sigma^2)^{-(n-1)/2} e^{-(n-1)s^2/2\sigma^2}. \qquad (7.2)$$

This is not all of the evidence about σ^2 in the complete sample $x_1, \ldots, x_n$, but it is, in various senses, most of it, and it is all of the evidence that can be disentangled from the nuisance parameter μ without additional assumptions.

The marginal likelihood function, (7.2), is not *the* likelihood function for σ^2 in the way that the orthogonal likelihood factor in (7.1) is the likelihood function for the ratio of Poisson means, $\lambda_1/\lambda_2 = \theta$. This marginal likelihood function might represent the best, most useful, one-dimensional summary of the evidence about σ^2 in this sample. But marginal likelihoods are not available in all problems where orthogonal reparameterization is impossible. And even when they are available, there are other candidates to consider for representing the evidence about a single parameter of interest in the presence of nuisance parameters. In addition to marginal likelihoods, there are conditional, estimated, profile, partial, and integrated (Bayesian) likelihoods, as well as many variations and hybrids, such as modified, adjusted, and conditional profile likelihoods. In this chapter we will consider some aspects of the first three of these and introduce one more, synthetic conditional likelihoods. Bayesian integrated likelihoods are discussed briefly in the next chapter.

Before turning to the alternatives, we should note the importance of marginal likelihoods – the procedure of using a marginal probability distribution model for some well-chosen function of the data instead of a full parametric model for the entire data vector is the key to many practical applications of likelihood inference. Not only does this procedure enable us, in some special cases, to avoid nuisance parameters that appear in our model for the entire data vector, as it did in the above example where the marginal model for the sample variance depended only on σ^2, and thus avoided the nuisance parameter θ in the $N(\theta, \sigma^2)$ model for the entire data vector. It also enables us, in a very wide class of applications, to avoid specifying a detailed model for the entire observation vector in the first place.

For example, when $X_1, \ldots, X_n$ are i.i.d. random variables, unless n is small, we need not specify the precise form of the distribution of a single element, X, in order confidently to model the marginal distribution of $\bar{X}$ as normal (because of the central limit theorem). When the variance of this marginal distribution is replaced by a consistent estimator, the resulting estimated likelihood function for the mean, $\mathrm{E}X$, is valid, in a specific approximate sense, under a wide range of specific parametric models for the distribution of X (Tsou and Royall, 1995).

Similar arguments can be used in support of normal marginal distribution models for a vast array of asymptotically normal estimators, including those generated by least squares, generalized linear models, estimating equations, etc. These marginal models for the distributions of estimators generate likelihood functions for the corresponding parameters. Thus the same models that provide the standard approximate confidence intervals for weighted least-squares estimators, for example, provide approximate likelihood functions for regression coefficients. And it is in these marginal distribution models that the possibility of orthogonal reparameterization of multivariate normal models finds its most important applications.

7.4 Conditional likelihoods

We have just seen that in the case of the normal variance, the marginal probability distribution of a well-chosen function of the data can provide clean evidence about a single parameter of interest, that is, evidence that can be represented and interpreted independently of nuisance parameters. Similar results can sometimes be obtained from conditional probability distribution models. An important example of this occurs in the case of the odds ratio for two independent binomial samples. If X and Y have independent $Bin(m, p_x)$ and $Bin(n, p_y)$ distributions, with m and n known, then the conditional distribution of X, given the total number of successes, $X + Y$, has a distribution that depends only on the odds ratio, $\psi = p_x(1 - p_y)/p_y(1 - p_x)$. This conditional distribution provides a likelihood function for ψ:

$$L_C(\psi) \propto \Pr(X = x | X + Y = x + y; p_x, p_y)$$

$$\propto \left[\sum \binom{m}{j} \binom{n}{x + y - j} \psi^{j-x}\right]^{-1}, \qquad (7.3)$$

where the lower limit of summation is $\max(0, x + y - n)$ and the upper limit is $\min(m, x + y)$. We saw examples of this conditional likelihood function in sections 6.3 and 6.7. Another example is Figure 7.1, which shows the evidence about the odds ratio in a data set used by Edwards (1972, p. 192). Of 121 families containing one set of same-sex twins, 56 had twin boys and 65 had twin girls. The relationship between the gender of the twins and the age of their mother at the birth of her first child is of interest. When mothers are divided into two groups according to that age, we have Table 7.1.

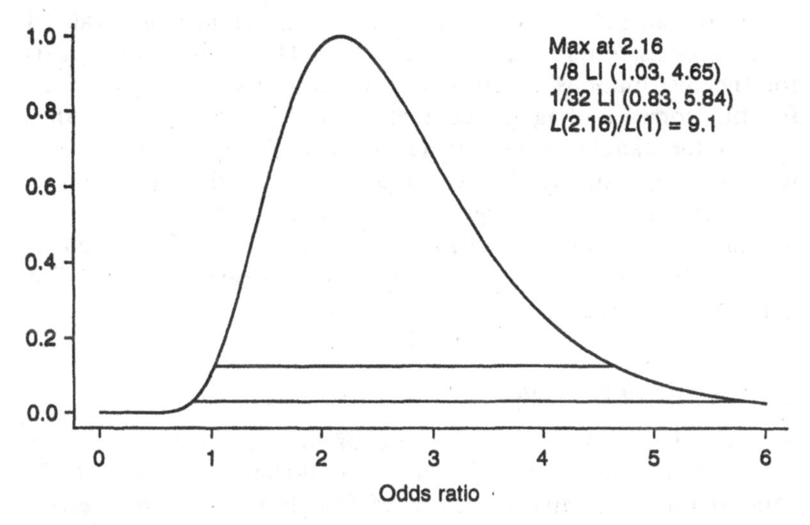

Figure 7.1 *Conditional likelihood for ratio of odds of female twins: younger versus older mothers.*

Table 7.1 *Age of mother at birth of first child*

Twin type	16–26	27–42	
FF	38	27	65
MM	22	34	56
	60	61	

Suppose these data were obtained by fixing the numbers of mothers in the two age groups (at 60 and 61), then determining for each mother the gender of her twins. In that case, we might use a model that treats these data as realizations of independent binomial variables, $Bin(60, p_y)$, and $Bin(61, p_o)$, where p_y is the probability that the twins' gender is female when the mother is in the younger (16–26) age group and p_o is the probability that the twins are female when the mother is in the older group. The evidence about the odds ratio, $p_y(1-p_o)/p_o(1-p_y)$, that is obtained from the conditional distribution, given that 65 sets of female twins are observed, is shown in Figure 7.1. An odds ratio of 1 corresponds to $p_o = p_y$, or no difference. These data are fairly strong evidence supporting an odds ratio of about 2 versus a ratio of 1, indicating a greater probability of female twins for a mother in the younger age group than for a mother in the older. In particular, the best-supported value of the odds ratio, 2.16, is better supported than 1 by a factor of 9.1. The 1/8 support interval barely excludes the values 1 and 5.

If a different sampling scheme had been used, with the number of female twins fixed at 65 and the number of male twins at 56, we might model these data differently, treating the number of mothers of female twins in the younger age group as a $Bin(65, p_F)$ variable, and the number of mothers of male twins in that age group as $Bin(56, p_M)$. Given that the total number of mothers in the younger age group is 60, the likelihood function for the odds ratio, $p_F(1-p_M)/p_M(1-p_F)$, under the new model is identical to that for the odds ratio under the first model. The odds ratio of 2 now indicates a greater probability that the mother is in the younger age group if she has female twins than if she has males. It is easy to show that the equality of the likelihood functions for the odds ratios under these two models for the data in Table 7.1 is no accident – the conditional likelihood function for the odds ratio is always the same when the rows of a 2×2 table of counts are modelled as independent binomial random variables as when the columns are modelled that way (Exercise 7.8).

7.5 Estimated likelihoods

Two other techniques eliminate nuisance parameters from a likelihood function by simply replacing them with estimates. The first of these treats the likelihood function with the nuisance parameter fixed at its true value, $L(\theta, \gamma_0)$, as a target, and estimates this function by $L(\theta, \hat{\gamma})$, where $\hat{\gamma}$ is a consistent estimator of γ, such as the MLE. This gives an 'estimated' likelihood for θ, $L_E(\theta) = L(\theta, \hat{\gamma})$.

Estimated likelihood functions have the interesting property that for any pair (θ_1, θ_2) the estimated likelihood ratio $LR_E = L_E(\theta_1)/L_E(\theta_2)$ is asymptotically equivalent, in a certain sense, to the likelihood ratio, $LR_0 = L(\theta_1, \gamma_0)/L(\theta_2, \gamma_0)$, that would be the appropriate measure of the support for θ_1 versus θ_2 if the true value of γ were known. (The two ratios are equivalent in the sense that their logarithmic mean difference, $[\ln(LR_E) - \ln(LR_0)]/n$, converges to zero almost surely (Tsou and Royall, 1995).)

The estimated likelihood function for the normal variance is the same as the marginal likelihood, expression (7.2), except that the 'degrees of freedom', $n-1$, in the first factor, $(\sigma^2)^{-(n-1)/2}$, is replaced by n. This illustrates a widely recognized weakness of estimated likelihood functions – they are 'too concentrated', that is to say, their shape is the same as if the value of the nuisance parameter were known to equal the estimated value. It seems intuitively clear that the fact that the nuisance parameter is not known should usually be reflected in a flatter likelihood function, signifying that the evidence about the parameter of interest is weaker than it is when the nuisance parameter's value is known.

7.6 Profile likelihoods

Another technique also removes γ by replacing it. But instead of using an estimate of the true value, γ_0, in place of γ, it uses, for each value of θ, the value, $\hat{\gamma}(\theta)$, that maximizes $L(\theta, \gamma)$ (Kalbfleisch and Sprott, 1970). This gives the 'profile' likelihood:

$$L_P(\theta) = L(\theta, \hat{\gamma}(\theta)) = \max_{\gamma}(L(\theta, \gamma)).$$

An important example of profile likelihoods occurs when the $N(\theta, \sigma^2)$ model is used, and σ^2 is a nuisance parameter. In this case the likelihood is maximized, for a fixed value of θ, by $\hat{\sigma}^2(\theta) = \sum(x-\theta)^2/n$, and the profile likelihood function is

$$L_P(\theta) = L(\theta, \hat{\sigma}^2(\theta)) \propto \left[\sum(x-\theta)^2\right]^{-n/2}.$$

A slightly modified version of this profile likelihood function, in which n is replaced by $n-2$, was used for the examples in sections 6.8 and 6.9.

For problems in which nuisance parameters cannot be eliminated by orthogonalization, marginal and conditional likelihoods sometimes provide satisfactory solutions. When these alternatives are not available, we can always turn to profile and estimated likelihoods. Marginal and conditional likelihoods have one important advantage over profile and estimated likelihoods, in general. Because they are 'real' likelihoods – that is, they are built from actual probability density or mass functions for observable random variables – general results about the probability of misleading evidence, such as the universal bound in section 1.4, apply to marginal and conditional likelihoods. Profile and estimated likelihoods do not share this important property, and for that reason, they must be used more cautiously. In particular, the universal bound on the probability of misleading evidence does not apply to profile likelihoods (Exercise 7.10).

One observation in favor of the profile likelihood is the following: if an orthogonalizing transformation of the parameter exists, so that an orthogonal likelihood for the parameter of interest exists, then the profile likelihood is that orthogonal likelihood. (The reasoning is simply that in an orthogonal parameterization the profile likelihood is obviously the orthogonal likelihood, and that the profile likelihood is unchanged by reparameterization (invariant) – see Tsou (1991).) That is, we need not actually find the orthogonal transformation. The profile procedure generates the orthogonal likelihood directly, regardless of whether the model is parameterized so that the nuisance parameter is orthogonal to the parameter of interest or not (Exercises 7.2 and 7.4).

7.7 Synthetic conditional likelihoods

Figure 7.1 showed a likelihood function for the odds ratio $\psi = p_x(1-p_y)/p_y(1-p_x)$. The full likelihood function for the two-binomial distribution model used in that example is given by

$$L(p_x, p_y) \propto p_x^x(1-p_x)^{m-x}p_y^y(1-p_y)^{n-y}, \tag{7.4}$$

and the likelihood function for the odds ratio was obtained from the conditional distribution of X, given $X+Y$ (expression (7.3)). Thus we used expression (7.3) to represent the main component (Plackett,

1977) of the evidence about ψ, evidence that is represented in full by (7.4).

We might just as well use the same representation in other cases where the full likelihood is given by (7.4), regardless of whether that likelihood was actually generated under a two-binomial probability model, or under some other model, such as two negative binomials. When the model is not the two-binomial, expression (7.3) is no longer a conditional likelihood. In such cases we will call it a 'synthetic conditional likelihood'.

The argument is as follows: a likelihood of the form (7.4), such as

$$p_x^{38}(1-p_x)^{22}p_y^{27}(1-p_y)^{34}, \tag{7.5}$$

can arise in more than one way. It can be generated by fixing the total numbers of Bernoulli trials in the two groups, 60 and 61, then observing $x = 38$ and $y = 27$ successes (the two-binomial model). It can also be generated by observing the number of trials required to obtain 22 failures in the first group and the number required to obtain 34 failures in the second, and observing 60 and 61 respectively (a two-negative-binomial model). The likelihood principle assures us that the evidence about p_x, p_y (and ψ) is represented by (7.5), and is the same in both cases. Formula (7.3) provides no more or less adequate representation of that evidence in one case than in the other. It is no less adequate in the negative binomial case, where it is a synthetic conditional likelihood, than in the binomial case, where it is a true conditional likelihood.

This argument enables us to determine, in the important two-binomial case, a synthetic conditional likelihood for the probability ratio, $\theta = p_x/p_y$, often called the 'relative risk' in epidemiology. This is the likelihood function shown in Figure 6.11. In the example at hand, if the observations were actually generated by making independent Bernoulli trials until 22 failures occurred in the first group (and finding that 60 trials were required) and 34 failures occurred in the second (requiring 61 trials) then, given that a total of 65 successes were observed, the conditional probability distribution depends only on the probability ratio θ, giving the conditional likelihood function (Exercise 7.12)

$$L_C(\theta) \propto \left[\sum_{j=22}^{87} \binom{j-1}{21}\binom{120-j}{33}\theta^{j-65}\right]^{-1}$$

This is the synthetic conditional likelihood for θ under the two-binomial model.

7.8 Summary

Some multiparameter models are structured so that the likelihood function for a single parameter of interest can be extracted. This is true of the two-Poisson model in section 7.2, where orthogonal likelihoods for the ratio of the two parameters and for their sum can be determined. In other cases orthogonal reparameterization is impossible, and then in order to satisfy our analytic desire to consider only one variable at a time we must resort to *ad hoc* methods or approximations.

As we saw in Chapter 6, these *ad hoc* methods, involving marginal, conditional, profile, and even synthetic likelihoods, appear to provide quite satisfactory results in most problems. The most promising general approach is that of profile likelihoods, perhaps modified along the lines suggested by Kalbfleisch and Sprott (1970), Cox and Reid (1987), Barndorff-Nielsen (1986), McCullagh and Tibshirani (1990) and others. However, the theoretical work that has been done on modifications to profile likelihoods has been directed mainly at the goal of producing point estimates and confidence intervals with good frequentist properties, and much remains to be understood in relation to the evidential interpretation of data via the likelihood paradigm.

Exercises

7.1 (a) Suppose counts x_1 and x_2 are modelled as realizations of independent Poisson random variables X_1 and X_2 with parameters λ_1 and λ_2. Show that the marginal likelihood function for $\theta_1 = \lambda_1 + \lambda_2$ based on the observation $X_1 + X_2 = x_1 + x_2$, that is, $L(\theta_1) \propto \theta_1^{x_1+x_2} \mathrm{e}^{-\theta_1}$, is an orthogonal likelihood. That is, show that this likelihood function can be obtained from the full data model by reparameterizing in terms of a new parameter (θ_1, θ_2) where θ_1 and θ_2 are orthogonal. [*Hint*: $\theta_2 = \lambda_1/(\lambda_1 + \lambda_2)$ will work.]

(b) Generalize the result in (a) to more than two λs. If X_i are independent *Poisson*(λ_i), $i = 1, \ldots, n$, the likelihood function for $\sum \lambda_i$ based on the observation $\sum X_i = \sum x_i$ is an orthogonal likelihood function. (From this it follows that an orthogonal likelihood for any sum of the λ_i can be similarly obtained from the Poisson distribution model for the sum of the corresponding X_i.) It can be obtained from the joint likelihood for $(\lambda_1, \ldots, \lambda_n)$ based on

$(X_1, \ldots, X_n) = (x_1, \ldots, x_n)$ by reparameterization in terms of $(\theta_1, \theta_2, \ldots, \theta_n)$, where $\theta_1 = \sum \lambda_i$ is orthogonal to $(\theta_2, \ldots, \theta_n)$.

(c) Show that in the case $n = 2$, an orthogonal likelihood function for the ratio $\theta = \lambda_1/\lambda_2$ is $L(\theta) \propto \theta^{x_1}/(1+\theta)^{x_1+x_2}$, $0 < \theta < \infty$.

7.2 (a) (Continuation of Exercise 7.1(a)). Suppose we reparameterize in terms of $\theta_1 = \lambda_1 + \lambda_2$ and λ_2. (Note that θ_1 and λ_2 are not orthogonal.) Show directly that the profile likelihood function for θ_1 is the orthogonal likelihood given in Exercise 7.1(a).

(b) Reparameterize in terms of $\theta_2 = \lambda_1/\lambda_2$ and λ_2. Show directly that the profile likelihood function for θ_2 is the orthogonal likelihood given in Exercise 7.1(c).

7.3 Suppose X has a multivariate normal distribution with mean vector μ and fixed covariance matrix Σ. Suppose μ is partitioned $\mu = (\mu_1, \mu_2)$ where μ_1 is a parameter vector of interest; let (X_1, X_2) be the corresponding partition of X. Show that:

(a) this model can be reparameterized in terms of a mean vector $\theta = (\theta_1, \theta_2)$, where θ_1 equals μ_1 and is orthogonal to θ_2;

(b) the orthogonal likelihood for $\mu_1 = \theta_1$ is the one obtained from the marginal distribution of X_1,

$$L(\mu_1, x) \propto \exp\{(x_1 - \mu_1)'\Sigma_{11}^{-1}(x_1 - \mu_1)/2\}.$$

where Σ_{11} is the covariance matrix of X_1. [*Hint*: The required reparameterization is given by $\theta = A\mu$, where

$$A = \begin{bmatrix} I & 0 \\ -\Sigma_{21}\Sigma_{11}^{-1} & I \end{bmatrix}.]$$

7.4 (This exercise requires some familiarity with standard results on minimizing quadratic forms and inverting partitioned matrices.) Show that in the original parameterization of Exercise 7.3, $\mu = (\mu_1, \mu_2)$, the profile likelihood function for μ_1 is the orthogonal likelihood given in Exercise 7.3(b).

7.5 Suppose the vector-valued random variable X has a multinomial probability distribution with cell probabilities $(\pi_1, \pi_2, \ldots, \pi_p)$, that is,

$$\Pr(X = (x_1, \ldots, x_p)) = \binom{n}{x_1, \ldots, x_p} \prod_{i=1}^{p} (\pi_i)^{x_i},$$

where $\sum x_i = n$ and $\sum \pi_i = 1$. Show that the orthogonal likelihood for $\theta = \pi_1 + \pi_2$ is given by $L(\theta) \propto \theta^{x_1+x_2}(1-\theta)^{n-x_1-x_2}$. Show the analogous result for $\theta = \sum' \theta_i$ where $\sum'$ represents summation over an arbitrary subset of the integers $\{1, 2, \ldots, p\}$.

7.6 Let X and Y be independent normal random variables with respective means (μ, η) and known variances (σ^2, τ^2).

(a) Show that the profile likelihood for the ratio of means, $\eta/\mu = \theta$, is

$$L_P(\theta) = \exp\{-\tfrac{1}{2}(y - \theta x)^2/(\tau^2 + \theta^2\sigma^2)\}.$$

(b) Show that the $1/k$ likelihood set is a finite interval when $x^2 > 2\sigma^2 \ln k$, but that when this inequality is reversed, the set is either the entire real line or the entire line minus a finite interval.

(c) Show that the usual 95% confidence set derived from the fact that $(Y - \theta X)/\sqrt{\tau^2 + \theta^2\sigma^2}$ has a standard normal distribution is the likelihood set (b) with $k = \exp\{(1.96)^2/2\} = 6.826$.

7.7 Show that for the i.i.d. $N(\mu, \sigma^2)$ model for observations $x_1, \ldots, x_n$, the conditional distribution of $X_1, X_2, \ldots, X_n$, given the mean $\bar{X}$, provides a conditional likelihood function for σ^2 that is identical to the marginal likelihood function given by expression (7.2).

7.8 (a) Consider four counts, a, b, c, and d, arranged in a 2×2 table

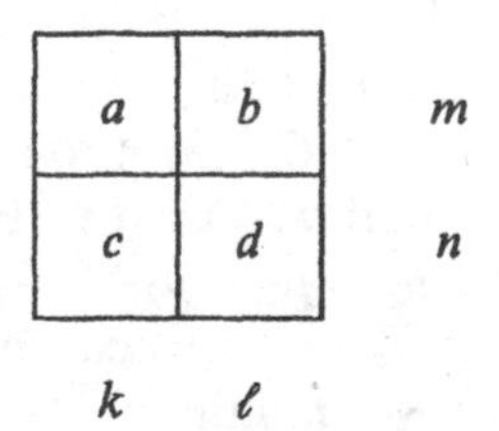

A study where two groups of subjects (group sizes m and n) are followed, with the incidence of a certain disease in each group noted (incidences a and c), produces data of this form. In this case the counts a and c might be modelled as realizations of independent binomial random variables, A and C, with parameters (m, ρ_1) and (n, ρ_2). The odds ratio in favor of catching the disease (in the first group versus the second) is $\psi_r = \rho_1(1-\rho_2)/\rho_2(1-\rho_1)$. Use the conditional distribution of A, given the total number of disease cases observed, $A + C = k$, to find the conditional likelihood function for ψ_r.

(b) The same table might also be produced by a 'case-control' study, in which k cases of the disease and ℓ controls are classified according to whether they belong to group 1 or group 2 (e.g. those exposed to some risk factor and those not exposed). A two-binomial model might be used in this case also, but now it is a and b that are treated as realizations of independent binomial variables, A and B, with parameters (k, γ_1) and (ℓ, γ_2). The odds ratio in favor of belonging to group 1 (in cases versus controls) is $\psi_c = \gamma_1(1 - \gamma_2)/\gamma_2(1 - \gamma_1)$. Use the conditional distribution of A, given $A + B = m$, to find a likelihood function for ψ_c. Compare this to the likelihood function for ψ_r found in part (a).

7.9 In Exercise 7.8 we showed that a given 2×2 table generates the same conditional likelihood function for the odds ratio ψ_r under a two-binomial model for the rows as for the odds ratio ψ_c under a two-binomial model for the columns. Consider a multinomial model that treats the four counts (a, b, c, d) as realizations of a random vector (A, B, C, D) with

$$\Pr((A, B, C, D) = (a, b, c, d)) = \frac{(a + b + c + d)!}{a!b!c!d!} p_{11}^a p_{12}^b p_{21}^c p_{22}^d,$$

where p_{ij} is the probability of the cell in row i and column j, with $p_{11} + p_{12} + p_{21} + p_{22} = 1$. Show that the conditional distribution, given $A + B = m$ and $A + C = n$, produces the two-binomial model for the rows that was used in Exercise 7.8(a), with $\rho_1 = p_{11}/(p_{11} + p_{12})$ and $\rho_2 = p_{21}/(p_{21} + p_{22})$.

Show that the model in Exercise 7.8(b) can be derived in a similar way, and that under the multinomial model the two odds ratios ψ_r and ψ_c are actually the same parameter,

$$\psi = p_{11}p_{22}/p_{12}p_{21}.$$

7.10 Use the $N(\mu, \sigma^2)$ model to show that the universal bound on the probability of misleading evidence does not apply to profile likelihoods. [*Hint*: Let μ be the nuisance parameter, and consider the extreme case of $n = 1$.]

7.11 Consider the $N(\mu, \sigma^2)$ model for a single observation when σ^2 is a nuisance parameter. Show that the probability that the profile likelihood ratio $L_P(\mu_2)/L_P(\mu_1)$ will exceed 8 when $\mu = \mu_1$ is maximized at $\mu_2 = \mu_1 + \frac{63}{32}(\ln\frac{9}{7})^{1/2}\sigma$, and that the maximum probability is 0.0605.

7.12 Suppose the 2×2 table in Exercise 7.8 arises by fixing b and d, then observing (i) m, the number of i.i.d. *Bernoulli*(p_x) trials required to produce b failures and (ii) n, the number of i.i.d. *Bernoulli*(p_y) trials required to produce d failures. Show that the conditional probability distribution of m, given $m+n$, depends only on the probability ratio $\gamma = p_x/p_y$, and that this conditional likelihood function for γ is

$$L_C(\gamma) \propto \left[\sum_{j=b}^{m+n-d} \binom{j-1}{b-1}\binom{m+n-j-1}{d-1}\gamma^{j-m}\right]^{-1}.$$

CHAPTER 8

Bayesian statistical inference

8.1 Introduction

Contemporary statistical practice consists largely of an informal mixture of the Neyman–Pearson and Fisherian concepts and methods that were described in Chapters 2 and 3. These are generally referred to as 'frequentist' in discussions that distinguish between them and a radically different body of concepts and methods that are called 'Bayesian'. Bayesian statistics is fundamentally different from frequentist statistics, and the strengths and weaknesses of the two schools have been the subject of intense debate. Thus the most closely studied division within statistics is not the one between the Neyman–Pearson and Fisherian schools of frequentists; it is the division between the frequentists, of whatever persuasion, and the Bayesians (see, for example, Lindley, 1975; Efron, 1978; 1986; Breslow, 1990).

Bayesian statistical methods are distinctive at every level. They differ from frequentist methods with respect to the formal probability models that are used, the meaning associated with the probabilities in the models, how the variables within the models are manipulated in analyses, and the results.

8.2 Bayesian statistical models

Bayesian and frequentist probability models have one critical, and conspicuous, difference. Here is a simple example. I have glued together a quarter, a nickel, and a dime to make a coin of new denomination, a 40c piece. One side of this coin is the 'tails' side of the quarter, and the other is the 'heads' side of the dime. I intend to toss this coin repeatedly and to observe whether it falls heads or tails. Statisticians of all persuasions will comfortably represent the results of the tosses as independent random variables $X_1, X_2, \ldots, X_n$ taking values one (heads) and zero (tails), with a common probability, θ, of heads. What values $x_1, \ldots, x_n$ will be observed is uncertain, and the parameter θ is fixed but unknown. The frequentist's modelling process stops here. But for the Bayesian approach, this model is inadequate; another component must be added.

The Bayesian approach requires that θ too must be represented as a realized value of a random variable, Θ, so that to the frequentist's model must be added a probability density for Θ. It is this explicit requirement that the parameter of interest must be given a probability distribution that characterizes Bayesian statistical models, and it is here that the Bayesians and the frequentists part ways. The distribution for Θ is called the **prior** probability distribution, signifying that it represents the uncertainty about the value of θ before the random variable $X = (X_1, \ldots, X_n)$ is observed. We denote the density function of this prior probability distribution by $f_\Theta(\theta)$. In the Bayesian model the frequentist's probability density for X, for fixed θ, $f(x;\theta)$, is the conditional density, given $\Theta = \theta$, and is denoted by $f_{X|\Theta}(x|\theta)$.

The model for the observable random variable X, that is, the part of the model that the frequentists and Bayesians share, consists of a family of probability distributions indexed by the parameter θ. Because the value of θ is unknown, the distribution of X is unknown. The Bayesian model for the distribution of the parameter Θ is not of this sort – $f_\Theta(\theta)$ is a single, completely specified probability density function. In the example of the 40c piece, if the Bayesian uses the beta probability model with parameters (a, b) to represent the prior distribution of the probability of heads, the values of (a, b) must be specified. (If the values are not specified and (a, b) are instead left as variables, then they also must be represented as random variables and given their own prior probability distribution. Thus the three parameters (θ, a, b) are given a joint prior probability model from which a and b can be integrated to yield the single, completely specified marginal prior distribution for Θ.)

The requirement that the Bayesian statistical model must include a prior probability distribution for θ is a direct result of the way the Bayesian formulates the problem of statistical inference. In that formulation the objective is to determine the probability distribution of the parameter, given the observations $x_1, \ldots, x_n$. And the reason why this school of statistics is called 'Bayesian' is that Bayes's theorem shows how to solve the problem:

$$f_{\Theta|X}(\theta|x) = c f_{X|\Theta}(x|\theta) f_\Theta(\theta), \tag{8.1}$$

where the constant $c = \int f_{X|\Theta}(x|\theta) f_\Theta(\theta)\,\mathrm{d}\theta$. The Bayesian solution to the problem of statistical inference is the **posterior** probability distribution of θ, given $X = x$, represented by the density function $f_{\Theta|X}(\theta|x)$. It cannot be determined from the frequentist's model, which, in Bayesian notation, supplies only the component $f_{X|\Theta}(x|\theta)$. Bayes's theorem (8.1) shows that for determining the posterior distribution another component, the prior density, $f_{\Theta}(\theta)$, is also essential.

8.3 Subjectivity in Bayesian models

Let us look more closely at the prior probability distribution for the parameter Θ when my 40c piece is tossed. Suppose for simplicity that the Bayesian uses the beta distribution with parameters $(a, b) = (1, 1)$, so that Θ has a *Uniform*$(0, 1)$ prior distribution. Using this distribution, he can calculate the probability that Θ is less than 1/3, and find that it equals 1/3. He can also discover that the probability that Θ falls in the interval $(0.5, 0.6)$ is 1/10, etc.

The Bayesian can not only calculate the probability that Θ will fall within any specified interval, but also calculate the probability that the first toss of the 40c piece will fall heads. It equals $E\Theta = 1/2$. He can determine that the probability that seven of the first ten tosses will fall heads is $120 \int_0^1 \theta^7(1 - \theta)^3 \, d\theta = 0.091$, and discover that this is precisely the same as the probability that none of the first ten tosses will fall heads. On the other hand, if he specifies that $(a, b) = (6, 3)$, this will imply that the probability of seven heads in the first ten tosses is 0.181, and that the probability of none is less than 0.002.

How does the Bayesian determine which prior to use? How can he know that the probability that the first ten tosses of the first 40c piece on Earth will produce seven heads is 0.181, and not 0.091, or some other value? If two Bayesians choose different prior probability distributions for Θ, how can anyone determine which one, if either, is correct? To answer these questions, we must ask another: What are these prior probabilities supposed to represent? The answer to this question explains why, despite the widely acknowledged shortcomings of frequentist statistical methods such as hypothesis tests and confidence intervals, Bayesian statistical methods have not replaced them. The answer is this: the prior probability distribution is supposed to represent the user's personal state of uncertainty about θ. If you and I bring different beliefs about θ to the problem, and if these beliefs are correctly represented by different prior probability distributions, then both distributions are correct. Yours is correct for you, and mine is for me. If, before any tosses are made, your beliefs about θ are correctly described by the *Uniform*(0, 1) probability distribution, then your probability that the first ten tosses will produce seven heads is 0.181.

Thus the Bayesian formulation of the problem of statistical inference leads to a model with an essential component that Bayesians acknowledge to be frankly subjective. Edwards, Lindman, and Savage (1963), in a profound and influential presentation of the Bayesian view, state that 'the key ideas of Bayesian statistics [are] that probability is orderly opinion, and that inference from data is nothing other than the revision of such opinion in the light of relevant new information'. The Bayesian approach, they say, 'is simply a set of techniques for orderly expression and revision of your opinions with due regard for internal consistency among their various aspects and for the data'. Another leading Bayesian, Lindley (1965), explained that 'the main subject matter of statistics is the study of how data sets change degrees of belief; from prior, by observation of A, to posterior. They change by Bayes's theorem ...'.

The frank subjectivity of the prior probability distribution has been a distinct handicap to those who have sought to persuade scientists to replace frequentist statistical methods with Bayesian ones. Efron (1986) explained that one of 'the main reasons why Fisherian and NPW [Neyman–Pearson] ideas have shouldered Bayesian theory aside in statistical practice' is that 'The high road of scientific objectivity has been seized by the frequentists'.

But Bayesians point out that frequentist methods have their own subjective component, and it is one that Bayesian statistics avoids:

> even standard statistical methods turn out to be based on subjective input – input of a type that science should seek to avoid. In particular, standard methods depend on the intentions of the investigator, including intentions about data that might have been obtained but were not. (Berger and Berry, 1988, p. 159)

Edwards, Lindman, and Savage (1963, p. 239) were making the same point 25 years earlier: 'Classical procedures ... insist that the intentions of the experimenter are crucial to the interpretation of data.' What these Bayesian writers are pointing out is that the dependence of Neyman–Pearson and Fisherian methods on the sample space makes frequentists' interpretation of experimental data subjective. We have seen many examples of this in earlier chapters. In section 3.4 we observed that frequentist interpretation of my observation of six successes in ten independent Bernoulli trials depends on whether or not I would have consulted my code-book if a different number of successes had occurred. Another example is given in section 5.5, where the frequentists' measure of the strength of evidence against the hypothesis of no difference depends on what the investigator really would have done if the observed difference had not been in the direction that he expected.

The Bayesians are quite right. Current frequentist methodology does have its own element of subjectivity. Both Bayesian and likelihood methods are protected from that particular sort of subjectivity by the likelihood principle, and its implication, the irrelevance of the sample space. But the Bayesians, in their prior probability distributions, introduce another component of subjectivity, which Berger and Berry (1988) acknowledge and defend. Fortunately, we are not forced to choose either of these two evils, the sample-space dependence of the frequentists or the prior distributions of the Bayesians. Likelihood methods avoid both sources of subjectivity.

8.4 The trouble with Bayesian statistics

The trouble with Bayesian statistics, from the standpoint of representing and interpreting statistical data as evidence, is that it addresses the wrong question. We have seen that much of what is wrong with frequentist statistics is the result of using Neyman–Pearson methodology, which is directed to the question 'What should I do?', for answering the quite different question, 'What do these data say?'. In the same way, the main objections to Bayesian statistics spring from attempts to use Bayesian methodology for answering the latter question instead of the question for which that methodology is appropriate, namely, 'What should I believe?'.

Powerful arguments for consistency and coherence of beliefs support the Bayesian answer to the question about beliefs (Savage, 1954). And it is obvious that the answer to that question must depend, not only on the most recent set of observations, but also on what you believed before you made those observations; if the statistical evidence is weak then your posterior probability distribution will be determined mainly by your prior beliefs. The subjective component of the Bayesian model is an essential part of the correct answer to the Bayesian's question. Likewise, as we argued in Chapter 1, a reasonable answer to the question 'What should I do?' will depend not only on the statistical evidence from the present study, but also on prior beliefs and on what actions are available, as well as on a loss function, as represented in Bayesian decision theory (Cornfield, 1969; DeGroot, 1970). But what of the quite different and also important question that science relies on statistics to answer, 'What do these data say?'.

8.5 Are likelihood methods Bayesian?

We have argued that the correct answer to the problem of objectively representing and interpreting the evidential meaning of statistical observations is embodied in the law of likelihood and its consequence, the likelihood principle. In one important sense, this answer is a part of Bayesian statistics. Because the data x enter the Bayesian analysis only via the likelihood function, $f_{X|\Theta}(x|\theta)$, most Bayesians accept the likelihood principle (and Bayesians have been among its strongest advocates; see Edwards, Lindman, and Savage, 1963, p. 237; Basu, 1975; Lindley, 1982, p. 198; Berger and Wolpert, 1988). Moreover, some Bayesians have supported using the likelihood function *per se* as a vehicle for representing and communicating statistical evidence: 'If a Bayesian is a subjectivist he will know that the initial probability density varies from person to person and so he will see the value of graphing the likelihood function for communication' (Good, 1976).

However, in other important senses the likelihood concepts and methods discussed in this book are clearly non-Bayesian. The question that is answered is different, 'What do these data say?', not 'What do (or should) I believe?'. The probability model is different – likelihood analysis is based on only the non-controversial part of the Bayesian model, the part that the Bayesians and frequentists share. It does not depend on the Bayesian's subjective prior probability distribution. The form of the result is different – it is a likelihood function, which is a direct representation of the statistical evidence; it is not a posterior probability density function, which represents the uncertainty about the parameter via a synthesis of that evidence with prior opinion (section 1.13).

While Bayesians have embraced the likelihood principle, they have been less consistently enthusiastic about the law of likelihood. Berger and Wolpert (1988), for example, in a monograph that is 'essentially aimed at promoting the LP [likelihood principle]', do not mention the law of likelihood by name. They do, however, address the concept briefly, in their final chapter: 'Most of the likelihood methods that have been proposed rest on the interpretation that $[L(\theta_1; x)/L(\theta_2; x)]$ measures the relative support of the data for θ_1 and θ_2.' But they show little interest in this 'interpretation', choosing instead to 'concentrate on indicating how sensible use of the likelihood function seems possible only through Bayesian analysis'.

Lindley (1992, p. 415) even goes to the extreme of rejecting the law of likelihood. He suggests that Simpson's paradox reveals a fatal flaw in the law, and concludes that 'likelihood will not do'. Regardless of whether Lindley is convinced by the rebuttal of his argument (Royall, 1992, p. 417), it is clear that likelihood concepts and methods are not appropriately viewed as simply a part of contemporary Bayesian statistics.

8.6 Objective Bayesian inference

The Bayesian posterior probability distribution that is generated by statistical data x and a particular prior distribution for θ represents what one should believe about θ after seeing the data *if* one's beliefs before seeing the data were properly represented by that prior. A prior probability distribution that represents the absence of belief, one that represents perfect ignorance or indifference, the state of mind of an ideal ignoramous, would then produce a posterior probability distribution that represents what one should believe about θ on the basis of the evidence x alone.

With such a 'neutral' or 'non-informative' prior, the Bayesians would have an answer to our question, 'What do the data say?'. The posterior probability distribution derived from that prior would represent what the data say about θ. It would answer the question, and it would be free of the subjective component that has been the Achilles' heel of Bayesian statistics. Moreover, this answer, because it is a probability distribution, would have distinct advantages over the likelihood function as a representation of the statistical evidence. For one thing, the solution could be immediately appreciated by anyone who understands probabilities – likelihood concepts would not be needed. For another, because the operations of adding or integrating over sets of parameter values would be valid, composite hypotheses would present no special problem, and nuisance parameters could be eliminated so effortlessly that they would no longer deserve their pejorative name.

Naturally, Bayesians have devoted enormous efforts to the pursuit of non-informative priors, ones that would represent the absence of knowledge, belief, opinion, prejudice, or preference. These efforts have been unsuccessful for a simple reason – pure ignorance cannot be represented by a probability distribution.

Suppose, for example, that the parameter of interest is a probability θ, and that there is a probability distribution with density function $f(\theta)$ that correctly represents a state of pure ignorance about θ. This would be the appropriate prior to use for representing evidence about the probability of heads when I toss my 40c piece. Now consider another probability, γ, of which I am also ignorant, namely, the chance that the first two tosses of the coin will both fall heads. Because $\gamma = \theta^2$, the prior probability function for γ is $f(\gamma^{1/2})/2\gamma^{1/2}$. Thus if I give θ a neutral prior $f(\theta)$ then γ has a non-neutral prior. And if I give the second probability, γ, the neutral prior then θ has a non-neutral density, $2\theta f(\theta^2)$. If there is a probability density function that expresses ignorance, then I cannot be ignorant of both the probability of heads on one toss and the probability of heads on two consecutive tosses. As Edwards (1969, p. 1235) put it, in discussing the particular choice of the uniform prior, 'the uniform prior distribution used to convey complete ignorance about a parameter led to apparently gratuitious information about any new parameter functionally related to the first'.

The reason why pure ignorance cannot be represented by a probability distribution is that every probability distribution represents a particular state of uncertain knowledge; none represents the absence of knowledge. Knowing nothing about θ does not mean that the probability that θ is between 1/2 and 2/3 is 1/6, or that it is any other value. It means that I am unable to state a value for that probability. It is one thing to state that I do not know which of two possible values of θ is true, or that I have no knowledge or no prior evidence about which is true. It is quite another to assert that the two values are equally probable (Salmon, 1966, Chapter V; Edwards, 1972, section 4.5).

8.7 Bayesian integrated likelihoods

Bayesians attempt to proselytize frequentists who find likelihood concepts attractive, but who are frustrated by the lack of a definitive solution to the problem of nuisance parameters, by promising an easy solution. The Bayesians explain that, in their model, the likelihood function, $L(\theta, \gamma)$, is proportional to the conditional density at the observation, given the values of the random parameters, (θ, γ) – that is, it is proportional to $f(x|\theta, \gamma)$. The likelihood function for θ alone is then (proportional to) the conditional density, given θ alone. Under the Bayesian model it is obtained simply by 'integrating out' the nuisance parameter – that is, by integrating

$f(x, \gamma|\theta) = f(x|\theta, \gamma)g(\gamma|\theta)$ over the range of γ:

$$f(x|\theta) = \int f(x|\theta, \gamma)g(\gamma|\theta)\,d\gamma.$$

Here $g(\gamma|\theta)$ is the prior probability density of γ, given θ.

From a frequentist point of view, this is equivalent to using, not the original two-parameter model, $f(x|\theta, \gamma)$, but a new one-parameter model, $f(x|\theta)$, for the observable random variable X. By integrating out the nuisance parameter, the Bayesian replaces the two-parameter model $f(x|\theta, \gamma)$ for the probability distribution of X by a one-parameter model $f_g(x|\theta)$. Here we use the subscript, g, to emphasize that this model is determined by the prior distribution for the nuisance parameter, $g(\gamma|\theta)$. A Bayesian with a different prior for γ, say $g'(\gamma|\theta)$, will employ a different one-parameter model, $f_{g'}(x|\theta)$, for the observation.

For example, the Bayesian who uses the convenient *Gamma*(r, λ) prior probability distribution for $1/\sigma^2$ in an $N(\theta, \sigma^2)$ model replaces the two-parameter normal distribution model by a one-parameter model where the observable n-vector X has density

$$f(x|\theta; r, \lambda) = \frac{\Gamma((n/2) + r)}{\pi^{n/2}\Gamma(r)(\sum(x_i - \theta)^2 + 2\lambda)^{(n/2)+r}}. \qquad (8.2)$$

Here r and λ are specified values of the parameters in the gamma prior distribution, and the random variables $X_1, X_2, \ldots, X_n$ are no longer independent.

The Bayesian can remove the nuisance parameter from the likelihood because he has added another element to the model, the prior probability distribution. In an important sense, the frequentist was better off with the nuisance parameter. This is because, unlike the prior, the nuisance parameter is empirically meaningful – the data represent objective evidence about its value.

8.8 Summary

Frequentist statistical analysis treats the observations as realized values of random variables. The analysis is carried out in terms of a probability model for the distribution of these random variables. Bayesian analysis treats not only the observations, but also the objects of inference, the unobserved parameters, as random. It requires that another component be added to the probability model, the (prior) probability distribution of the parameters. The prior probability distribution represents the uncertainty about the parameters before the observations are made. In general, this uncertainty is essentially subjective.

The goal of Bayesian analysis is to represent the uncertainty about the parameters after the observations have been made. This uncertainty is quantified in the posterior probability distribution, which is obtained by using the likelihood function to modify the prior distribution, as directed by Bayes's theorem.

Efforts to interpret a particular posterior probability distribution as representing the evidence in the observations about the parameters fail for a simple reason: every prior distribution represents a specific state of knowledge, and none represents pure ignorance, so that the posterior distribution represents, not just the evidence in the data, but a synthesis of that evidence and prior knowledge. The evidence in the observations is represented by the likelihood function, and is measured by likelihood ratios. It is not represented and measured by probabilities, either frequentist sample-space probabilities or Bayesian posterior probabilities.

Exercises

8.1 For the $N(\theta, \sigma^2)$ model, show that if the prior distribution for σ^2 is such that $\gamma = 1/\sigma^2$ has a *Gamma*(r, λ) distribution, with density function

$$g(\gamma) = \lambda^r \gamma^{r-1} e^{-\lambda\gamma}/\Gamma(r), \qquad 0 < \gamma < \infty,$$

then the likelihood function for θ is given by expression (8.2).

8.2 (Continuation of 8.1.) The parameters (r, λ) must be given positive values in order to determine the prior probability distribution. The values $(0, 0)$, when substituted in the gamma density formula, determine a function that is not integrable, and is therefore not a probability density function. It is sometimes used anyway, and called an 'improper' prior density function. Show that with $(r, \lambda) = (0, 0)$, expression (8.2) is the profile likelihood for θ.

APPENDIX

The paradox of the ravens

Our problem, the interpretation of statistical data as evidence, is part of the more general philosophical problem of induction. Hempel's 'paradox of the ravens' has done much to stimulate and clarify thinking about the general problem. Let us see what light this paradox sheds on our special case and vice versa.

Hempel's example concerns a single hypothesis A which implies that under certain conditions a specified result x will be observed. Of course, if something else, not x, is seen, that represents decisive evidence against A, and A is disproved. But what of the case when x is observed, just as A predicted? Is this properly interpreted as evidence supporting A? In Hempel's famous example, A is the hypothesis 'All ravens are black', which implies that whenever a raven is observed, it will be found to be black. The suggested evidential interpretation – that the observation of a black raven is evidence supporting the hypothesis – seems reasonable. Hempel pointed out, however, that A is logically equivalent to another hypothesis, A', stating that 'All non-black objects are non-ravens'. Now if observation of a raven that is black is evidence supporting A, then observation of a non-black object that is not a raven is evidence supporting A'. But since A and A' are equivalent, this means that observation of a non-black non-raven (a red pencil, for example) is evidence that all ravens are black. Something seems wrong here.

Implicit in the raven paradox are issues of population size and sampling procedure. It is instructive to reformulate the problem so that these are made explicit. Let the two classes of objects, ravens and non-ravens, be represented by balls in two urns, R and NR; the balls in urn R are marked with the letter R, indicating that they are 'ravens', and those in NR are marked NR to show that they are the 'non-ravens'. All black objects are represented by black balls, and non-black objects by white ones. Now A is the hypothesis that the proportion of the balls in the raven urn (R) that are black, p_R, equals one: $H_0: p_R = 1$. If we represent the result of drawing a ball at random from that urn by the random variable X, with possible values b (black) and w (white) then the observation $X = b$ is indeed evidence supporting A over an alternative hypothesis, say B, asserting that p_R is some fraction $p < 1$. This is because the likelihood ratio is $\Pr_A(X = b)/\Pr_B(X = b) = 1/p > 1$. And according to the law of likelihood, n independent random draws (with replacement), all producing black balls, support A over B by the factor $(1/p)^n$, so that for large n the observations are strong evidence for the hypothesis that all ravens are black (A) *vis-à-vis* the hypothesis that only a fraction p are black (B).

On the other hand, a randomly drawn ball from the other urn, NR (non-ravens), carries no information about A versus B, regardless of its color, because neither hypothesis, A or B, makes any prediction about a draw from NR. If Y represents a random draw from NR, then its distribution is the same under both hypotheses, so that $\Pr_A(Y = w)/\Pr_B(Y = w) = 1$. That is, observing a randomly selected non-raven to be non-black ($Y = w$) is not evidence (or is utterly neutral evidence) about A versus B. Why, then, did we reach a different conclusion in Hempel's version of the problem? It is still true that A, asserting that all balls in urn R are black, is equivalent to another hypothesis A', asserting that all of the white balls are in urn NR (non-ravens). But we are no longer swept along to Hempel's paradoxical conclusion that a white ball from NR (a non-black non-raven) is evidence supporting A. How that conclusion might be reached, and when it is in fact appropriate, is explained by changing the procedure for selecting and reporting the observations.

We have seen that if we simply draw a ball from R (i.e. observe X), then its color is relevant to A versus B, while if we draw from NR (observe Y), the color is irrelevant to these hypotheses. Now suppose that what is fixed initially is not the urn from which a ball is drawn, but the ball's color. Suppose, for example, that the urns are overturned, and that one ball is selected at random from among all the white ones. Suppose, furthermore, that the selected ball is found to have come from NR (i.e., it is found to be a non-raven). Again we have observed a white ball from urn NR (a non-black non-raven). But this time it *is* evidence that all of the balls in R are black.

To see this, look at the random variable Z representing the outcome of this draw: $Z = R$ if the selected (white) ball comes from urn R, and $Z = NR$ if it comes from NR. Now the probability that $Z = NR$ is the proportion of the white balls that come from

NR, $n_{\mathrm{NR}}(1-p_{\mathrm{NR}})/[n_{\mathrm{R}}(1-p_{\mathrm{R}})+n_{\mathrm{NR}}(1-p_{\mathrm{NR}})]$, where n_{R} and n_{NR} are the numbers of balls in the two urns, and p_{NR} is the proportion of balls in NR that are black. Hypothesis A asserts that $p_{\mathrm{R}}=1$, implying that Z *must* equal NR, $\Pr_A(Z=NR)=1$, while B implies that $\Pr_B(Z=NR)$ is less than 1. Thus the observation $Z=NR$ is evidence supporting A over B by

$$\Pr_A(Z=NR)/\Pr_B(Z=NR)=1+[n_{\mathrm{R}}(1-p_{\mathrm{R}})/n_{\mathrm{NR}}(1-p_{\mathrm{NR}})]>1.$$

Both $Y=w$ and $Z=NR$ represent the observation of a white ball from NR. But the distribution of the random variable Y is determined by what fraction of the balls in NY is black, so that $Y=w$ is evidence about that fraction. On the other hand, the distribution of Z is determined by what fraction of the white balls is in NR. And $Z=NR$ is evidence supporting A, which implies that the fraction equals one, over B, which implies that it has some smaller value.

We see that the observation of a red pencil *can* be evidence that all ravens are black. To make the proper interpretation, we must have an additional piece of information. Whether the observation is or is not evidence supporting the hypothesis (A) that all ravens are black versus the hypothesis (B) that only a fraction $p_{\mathrm{R}}<1$ are black is determined by the sampling procedure. A randomly selected pencil that proves to be red is *not* evidence that all ravens are black, but a randomly selected red object that proves to be a pencil *is*. In this latter case, the fact that there are vastly many more red objects than there are ravens (of whatever color), i.e. $n_{\mathrm{NR}}(1-p_{\mathrm{NR}})\gg n_{\mathrm{R}}$, means that the likelihood ratio in favor of A, which equals $1+[n_{\mathrm{R}}(1-p_{\mathrm{R}})/n_{\mathrm{NR}}(1-p_{\mathrm{NR}})]$, is only slightly greater than one, so the evidence is quite weak.

References

Anscombe, F.J. (1964) Normal likelihood functions, *Annals of the Institute of Statistical Mathematics*, **26**, 1–19.

Armitage, P. (1975) *Sequential Medical Trials* (2nd edn), New York: Wiley.

Armitage, P. and Berry, G. (1987) *Statistical Methods in Medical Research*, Oxford: Blackwell Scientific.

Bakan, D. (1970) The test of significance in psychological research, in D.E. Morrison and R.E. Henkel (eds) *The Significance Test Controversy*, Chicago: Aldine.

Barnard, G.A. (1949) Statistical Inference, *Journal of the Royal Statistical Society, Series B*, **11**, 115–149.

Barnard, G.A. (1967) The use of the likelihood function in statistical practice, in *Proceedings of the Fifth Berkeley Symposium on Mathematical Statistics and Probability*, Vol. 1 (eds L. LeCam and J. Neyman), Berkeley: University of California Press.

Barnard, G.A., Jenkins, G.M., and Winsten, C.B. (1962) Likelihood inference and time series (with discussion), *Journal of the Royal Statistical Society, Series A*, **125**, 321–372.

Barndorff-Nielsen, O.E. (1986) Inference on full or partial parameters based on the standardized log likelihood ratio, *Biometrika*, **73**, 307–322.

Bartlett, R.H., Roloff, D.W., Cornell, R.G., Andrews, A.F., Dillon, P.W., and Zwischenberger, J.B. (1985) Extracorporeal circulation in neonatal respiratory failure: a prospective randomized study, *Pediatrics*, **76**, 479–487.

Basu, D. (1975) Statistical information and likelihood (with discussion), *Sankhyā, Series A*, **37**, 1–71.

Begg, C.R. (1990) On inferences from Wei's biased coin design for clinical trials (with discussion), *Biometrika*, **77**, 467–484.

Berger, J.O. (1990) Birnbaum's theorem is correct: a reply to a claim by Joshi, *Journal of Statistical Planning and Inference*, **26**, 112–113.

Berger, J.O. and Berry, D.A. (1988) Statistical analysis and the illusion of objectivity, *American Scientist*, **76**, 159–165.

Berger, J.O. and Wolpert, R.L. (1988) *The Likelihood Principle* (2nd edn), Institute of Mathematical Statistics Lecture Notes – Monograph Series, Vol. 6 (ed. S.S. Gupta), Hayward, California: IMS.

Berkson, J. (1942) Tests of significance considered as evidence, *Journal of the American Statistical Association*, **37**, 325–335.

Berry, G. (1983) The analysis of mortality by the subject-years method, *Biometrics*, **39**, 173–184.

Birnbaum, A. (1962) On the foundations of statistical inference (with discussion), *Journal of the American Statistical Association*, **53**, 259–326.

Birnbaum, A. (1969). Concepts of statistical evidence, in *Philosophy, Science, and Method: Essays in Honor of Ernest Nagel* (eds S. Morgenbesser, P. Suppes and M. White), New York: St Martin's Press.

Birnbaum, A. (1972) More on concepts of statistical evidence, *Journal of the American Statistical Association*, **67**, 858–861.

Birnbaum, A. (1977) The Newman–Pearson theory as decision theory, and as inference theory; with a criticism of the Lindley–Savage argument for Bayesian theory, *Synthese*, **36**, 19–49.

Bjornstad, J.F. (1996) On the generalization of the likelihood function and the likelihood principle, *Journal of the American Statistical Association*, **91**, 791–806.

Box, G.E.P. (1980) Sampling and Bayes' inference in scientific modelling and robustness (with discussion), *Journal of the Royal Statistical Society, Series B*, **143**, 383–430.

Breslow, N. (1990) Biostatistics and Bayes, *Statistical Science*, **5**, 269–298.

Breslow, N.E. and Day, N.E. (1987) *Statistical Methods in Cancer Research, Volume II, The Design and Analysis of Cohort Studies*, Lyon, France: International Agency for Research on Cancer.

Burdette, W.J. and Gehan, E.A. (1970) *Planning and Analysis of Clinical Studies*, Springfield, Illinois: Charles C. Thomas.

Carnap, R. (1950) *Logical Foundations of Probability*, Chicago: University of Chicago Press.

Chernoff, H. and Moses, L.E. (1959) *Elementary Decision Theory*, New York: Wiley.

Clayton, D. and Hills, M. (1993) *Statistical Models in Epidemiology*, New York: Oxford University Press.

Cornfield, J. (1966) Sequential trials, sequential analysis, and the likelihood principle, *American Statistician*, **29**(2), 18–23.

Cornfield, J. (1969) The Bayesian outlook and its applications (with discussion), *Biometrics*, **25**, 617–657.

Cox, D.R. (1958) Some problems connected with statistical inference, *Annals of Mathematical Statistics*, **29**, 357–372.

Cox, D.R. (1977) The role of significance tests, *Scandinavian Journal of Statistics*, **4**, 49–70.

Cox, D.R. and Hinkley, D.V. (1974) *Theoretical Statistics*, London: Chapman & Hall.

Cox, D.R. and Reid, N. (1987) Parameter orthogonality and approximate conditional inference (with discussion), *Journal of the Royal Statistical Society, Series B*, **49**, 1–39.

Cox, D.R. and Snell, E.J. (1981) *Applied Statistics. Principles and Examples*, London: Chapman & Hall.

Daniel, W.W. (1991) *Biostatistics: A Foundation for Analysis in the Health Sciences* (5th edn), New York: Wiley.

Dar, R., Serlin, R.C., and Omer, H. (1994) Misuse of statistical tests in three decades of psychotherapy research, *Journal of Consulting and Clinical Psychology*, **62**, 75–82.

DeGroot, M.H. (1970) *Optimal Statistical Decisions*, New York: McGraw-Hill.

DeMets, D.L. (1987) Practical aspects of data monitoring: a brief review, *Statistics in Medicine*, **6**, 753–760.

Dixon, W.J. and Massey, F.J. (1969) *Introduction to Statistical Analysis* (3rd edn), New York: McGraw-Hill.

Dupont, W.D. (1983) Sequential stopping rules and sequentially adjusted *p*-values: Does one require the other? (with discussion), *Controlled Clinical Trials*, **4**, 3–35.

Durbin, J. (1970) On Birnbaum's theorem on the relation between sufficiency, conditionality, and likelihood, *Journal of the American Statistical Association*, **65**, 395–398.

Edwards, A.W.F. (1969) Statistical methods in scientific inference, *Nature*, **222**, 1233–1238.

Edwards, A.W.F. (1970) Likelihood (letter to the editor), *Nature*, **227**, 92.

Edwards, A.W.F. (1972) *Likelihood*, London: Cambridge University Press.

Edwards, W., Lindman, H., and Savage, L.J. (1963) Bayesian statistical inference for psychological research, *Psychological Review*, **70**, 450–499.

Efron, B. (1978) Controversies in the foundations of statistics, *American Mathematical Monthly*, **85**, 231–246.

Efron, B. (1986) Why isn't everyone a Bayesian? (with discussion), *American Statistician*, **40**, 1–11.

Fieller. E.C. (1954) Some problems in interval estimation, *Journal of the Royal Statistical Society, Series B*, **16**, 175–185.

Fisher, R.A. (1922) On the mathematical foundations of theoretical statistics, *Philosophical Transactions of the Royal Society, Series A*, **222**, 309–368.

Fisher, R.A. (1925) Theory of statistical estimation, *Proceedings of the Cambridge Philosophical Society*, **22**, 700–725.

Fisher, R.A. (1934) *Statistical Methods for Research Workers* (5th edn), London: Oliver and Boyd.

Fisher, R.A. (1956) *Statistical Methods and Scientific Inference*, Edinburgh: Oliver and Boyd.

Fisher, R.A. (1958) *Statistical Methods for Research Workers* (13th edn), New York: Hafner.

Fisher, R.A. (1959) *Statistical Methods and Scientific Inference* (2nd edn), New York: Hafner.

Fisher, R.A. (1966) *Design of Experiments* (8th edn), New York: Hafner.

Fisz, M. (1963) *Probability Theory and Mathematical Statistics* (3rd edn), New York: Wiley.

Gart, J.J. (1971) The comparison of proportions: A review of significance tests, confidence intervals and adjustments for stratification, *Review of the International Statistical Institute*, **39**, 149–169.

Gart, J.J. and Nam, J. (1988) Approximate interval estimation of the ratio of binomial parameters: A review and corrections for skewness, *Biometrics*, **44**, 323–338.

Good, I.J. (1962) Discussion of A. Birnbaum (1962), On the foundations of statistical inference, *Journal of the American Statistical Association*, **57**, 312.

Good, I.J. (1976) The Bayesian influence, or how to sweep subjectivism under the carpet, in *Foundations of Probability Theory, Statistical Inference, and Statistical Theories of Science*, Vol. II (eds W.L. Harper and C.A. Hooker), Dordrecht: Reidel.

Hacking, I. (1965) *Logic of Statistical Inference*, New York: Cambridge University Press.

Hacking, I. (1972) Review of Edwards (1972), *British Journal of the Philosophy of Science*, **23**, 132–137.

Hansel, C.E.M. (1966) *ESP: A Scientific Evaluation*, New York: Charles Scribner's & Sons.

Hill, B.M. (1973) Review of A.W.F. Edwards' *Likelihood*, *Journal of the American Statistical Association*, **68**, 487–489.

James, W. and Stein, C. (1961) Estimation with quadratic loss, in *Proceedings of the Fourth Berkeley Symposium on Mathematical Statistics and Probability*, Vol. 1 (ed. J. Neyman), Berkeley: University of California Press.

Joshi, V.M. (1990) Fallacy in the proof of Birnbaum's theorem, *Journal of Statistical Planning and Inference*, **26**, 111–112.

Kalbfleisch, J.D. (1975) Sufficiency and conditionality, *Biometrika*, **62**, 251–268.

Kalbfleisch, J.D. and Sprott, D.A. (1970) Application of likelihood methods to models involving large numbers of parameters (with discussion), *Journal of the Royal Statistical Society, Series B*, **32**, 175–208.

Lee, A.M. and Fraumeni, J.F. (1969) Arsenic and respiratory cancer in man: an occupational study, *Journal of the National Cancer Institute*, **42**, 1045–1052.

Lehmann, E.L. (1959) *Testing Statistical Hypotheses*, New York: Wiley.

Lehmann, E.L. (1993) The Fisher, Newman–Pearson theories of testing hypotheses: One theory or two? *Journal of the American Statistical Association*, **88**, 1242–1249.

Lindley, D.V. (1965) *Introduction to Probability and Statistics: Part I. Probability*, Cambridge: Cambridge University Press.

Lindley, D.V. (1975) The future of statistics – a Bayesian 21st century, *Advances in Applied Probability* (supp.), **7**, 106–115.

Lindley, D.V. (1982) Bayesian inference, in *Encyclopedia of Statistical Sciences*, Vol. 1 (eds S. Kotz and N.L. Johnson), New York: Wiley-Interscience.

Lindley, D.V. (1992) Discussion of Royall (1992) The elusive concept of statistical evidence, in *Bayesian Statistics*, Vol. 4 (eds J.M. Bernardo, J.O. Berger, A.P. David, and A.F.M. Smith), New York: Oxford University Press, pp. 415–416.

Lindley, D.V. and Scott, W.F. (1984) *New Cambridge Elementary Statistical Tables*, London: Cambridge University Press.

Louis, T.A. (1993) Review of *Medical Statistics: A Commonsense Approach*, M.J. Campbell and D. Machin, *Controlled Clinical Trials*, **14**, 251–252.

McCullagh, P. and Tibshirani, R. (1990) A simple method for the adjustment of profile likelihoods, *Journal of the Royal Statistical Society, Series B*, **52**, 325–344.

Morrison, D.E. and Henkel, R.E. (1970) *The Significance Test Controversy*, Chicago: Aldine.

Murphy, E.A. (1985) *A Companion to Medical Statistics*, Baltimore, MD: Johns Hopkins University Press.

Neyman, J. (1950) *First Course in Probability and Statistics*, New York: Henry Holt and Company.

Neyman, J. (1976) Tests of statistical hypotheses and their use in studies of natural phenomena, *Communications in Statistics – Theory and Methods*, **8**, 737–751.

Neyman, J. and Pearson, E.S. (1933) On the problem of the most efficient tests of statistical hypotheses, *Philosophical Transactions of the Royal Society, Series A*, **231**, 289–337.

Noether, G.E. (1971) *Introduction to Statistics*, Boston: Houghton Mifflin.

O'Rourke, P.P., Crone, R.K., Vacanti, J.P., Ware, J.H., Lillihel, C.W., Parad, R.B., and Epstein, M.F. (1989) Extracorporeal membrane oxygenation and conventional medical therapy in neonates with persistent pulmonary hypertension of the newborn: A prospective randomized study, *Pediatrics*, **84**, 957–963.

Pearson, E.S. (1938) 'Student' as a statistician, *Biometrika*, **38**, 210–250.

Peto, R., Pike, M.C., Armitage, P., Breslow, N.E., Cox, D.R., Howard, S.V., Mantel, N., McPherson, K., Peto, J., and Smith, P.G. (1976) Design and analysis of randomized clinical trials requiring prolonged observation of each patient, I: Introduction and design, *British Medical Journal*, **34**, 585–612.

Plackett, R.L. (1977) The marginal totals of a 2 × 2 table, *Biometrics*, **64**, 37–42.

Pratt, J.W. (1961) Review of Lehmann, E.L. (1959), *Testing Statistical Hypotheses, Journal of the American Statistical Association*, **56**, 163–166.

Putnam, H. (1974) The corroboration of theories, in *The Philosophy of Karl Popper* (ed. P.A. Schilpp), LaSalle, Illinois: Open Court.

Remington, R.D. and Schork, M.A. (1970) *Statistics with Application to the Biological and Health Sciences*, Englewood Cliffs, New Jersey: Prentice-Hall.

Robbins, H. (1970) Statistical methods related to the law of the iterated logarithm, *Annals of Mathematical Statistics*, **41**, 1397–1409.

Royall, R.M. (1986) The effect of sample size on the meaning of significance tests, *American Statistician*, **40**, 313–315.

Royall, R.M. (1991) Ethics and statistics in randomized clinical trials (with discussion), *Statistical Science*, **6**, 52–88.

Royall, R.M. (1992) The elusive concept of statistical evidence (with discussion), in *Bayesian Statistics*, Vol. 4 (eds J.M. Bernardo, J.O. Berger, A.P. David, and A.F.M. Smith), New York: Oxford University Press.

Royall, R.M. and Cumberland, W.G. (1981a) An empirical study of the ratio estimator and estimators of its variance (with discussion), *Journal of the American Statistical Association*, **76**, 66–68.

Royall, R.M. and Cumberland, W.G. (1981b) The finite-population linear regression estimator and estimators of its variance – an empirical study, *Journal of the American Statistical Association*, **76**, 924–930.

Royall, R.M. and Cumberland, W.G. (1985) Conditional coverage properties of finite population confidence intervals, *Journal of the American Statistical Association*, **80**, 355–359.

Salmon, W.C. (1966) *The Foundations of Scientific Inference*, Pittsburgh: University of Pittsburgh Press.

Salmon, W.C. (1983) Confirmation and relevance, in *The Concept of Evidence* (ed. P. Achinstein), Oxford: Oxford University Press, pp. 95–123.

Savage, L.J. (1954) *The Foundations of Statistics*, New York: Wiley.

Savage, L.J. (1962) *The Foundations of Statistical Inference* (*A Discussion*), London: Methuen.

Savage, L.J. (1970) Comments on a weakened principle of conditionality, *Journal of the American Statistical Association*, **65**, 399–401.

Sewell, W.H. and Shah, V.P. (1968) Social class, parental encouragement and educational aspirations, *American Journal of Sociology*, **73**, 559–572.

Siegel, S. and Castellan, N.J., Jr. (1988) *Nonparametric Statistics for the Behavioral Sciences*, New York: McGraw-Hill.

Snedecor, G.W. and Cochran, W.G. (1980) *Statistical Methods* (7th edn), Ames: Iowa State University Press.

Snell, E.J. (1987) *Applied Statistics. A Handbook of BMDP™ Analyses*, London: Chapman & Hall.

Sokal, R.R. and Rohlf, F.J. (1969) *Biometry*, San Francisco: W.H. Freeman.

Swan, T. (1993) Examples, in *The GLIM System. Release 4 Manual* (eds B. Francis, M. Green, and C. Payne), Oxford: Clarendon Press.

Tsou, T.-S. and Royall, R.M. (1995) Robust likelihoods, *Journal of the American Statistical Association*, **90**, 316–320.

Tsou, T.-S. (1991) Robust likelihoods, PhD dissertation, Johns Hopkins University, Baltimore, Maryland.

United Kingdom Collaborative ECMO Trial Group (1996) UK collaborative randomised trial of neonatal extracorporeal membrane oxygenation, *The Lancet*, **348**, 75–82.

Venn, J. (1876) *The Logic of Chance* (2nd edn), London: Macmillan.

Wald, A. (1939) Contributions to the theory of statistical estimation and testing hypotheses, *Annals of Mathematical Statistics*, **10**, 299–326.

Wald, A. (1950) *Statistical Decision Functions*, New York: Wiley.

Ware, J.H. (1989) Investigating therapies of potentially great benefit: ECMO (with discussion), *Statistical Science*, **4**, 298–340.

Ware, J.H. and Epstein, M.D. (1985) Comments on 'Extracorporeal circulation in neonatal respiratory failure: A prospective randomized study' by R.H. Bartlett *et al.*, *Pediatrics*, **76**, 849–851.

Watson, G.S. (1983) Hypothesis testing, in *Encyclopedia of Statistical Sciences*, Vol. 3 (eds S. Kotz and N.L. Johnson), New York: Wiley.

Index